T3-BUX-232

Sustainable Production, Life Cycle Engineering and Management

Series Editors

Christoph Herrmann
Sami Kara

For further volumes:
http://www.springer.com/series/10615

Sustainable Production, Life Cycle Engineering and Management

Modern production enables a high standard of living worldwide through products and services. Global responsibility requires a comprehensive integration of sustainable development fostered by new paradigms, innovative technologies, methods and tools as well as business models. Minimizing material and energy usage, adapting material and energy flows to better fit natural process capacities, and changing consumption behaviour are important aspects of future production. A life cycle perspective and an integrated economic, ecological and social evaluation are essential requirements in management and engineering. This series will focus on the issues and latest developments towards sustainability in production based on life cycle thinking.

The book series is affiliated to the Joint German-Australian Research Group "*Sustainable Manufacturing and Life Cycle Management*", www.sustainable-manufacturing.com

Foreword

Machine tools are essential elements within metalworking manufacturing systems. While quality and procurement costs are a primary concern to machine tool builders and users, little attention has been paid to the energy demand of these unit process systems. With increasing energy costs and awareness about the energy-related environmental implications, the pursuit for energy efficiency of machine tools is increasingly gaining importance in the past decade. An essential prerequisite for improving the energy efficiency is its measurement. A predefined minimum energy threshold enables the quantification of energy efficiency of a whole machine tool system and triggers the identification as well as initiation of adequate means for a holistic improvement.

However, a methodological evaluation of energy efficiency of machine tools and a structured support for realising energy efficiency improvements have been neglected so far. In this book, Mr. Zein presents an approach for energy performance management of machine tools by providing innovative methods and tools to quantify and evaluate the energy efficiency while guiding the improvement towards an unequivocal, minimum energy reference level. This concept is based on a methodological derivation of technically achievable energy limits and takes into account the machine tool and process design.

Based on the assessment of energy efficiency of machine tools, Mr. Zein additionally introduces means to guide and control the derivation of improvement measures for a structured transition towards energy efficient machine tools. The developed energy performance management provides for the first time a comprehensive, conceptual approach based on scientifically sound methods and tools. The proposed methodology also takes into account the industrial challenges and is suited to comply with these requirements.

The implementation of the developed concept is achieved by means of an empirical determination of energy limits for grinding machines. The potential for evaluating energy efficiency and the derivation of improvement measures are demonstrated by using the machine tools in the automotive industry as an example. The introduced energy performance management concept represents a vital step forward in the current state of research with an application-oriented scope. As a result it encourages progress in industry and in the research towards sustainability in production.

Prof. Christoph Herrmann
Technische Universität Braunschweig

Prof. Sami Kara
The University of New South Wales

Acknowledgments

The present book has been developed in the context of my work as a research associate within the Product and Life Cycle Management Research Group at the Technische Universität Braunschweig. First and foremost, I would like to express my sincerest gratitude to apl. Prof. Dr.-Ing. Christoph Herrmann, head of the research group. He provided me with excellent opportunities for research and enabled me to explore the challenges of life cycle engineering. His good advice, encouragement and support have been invaluable on both academic and personal level.

I thank Associate Professor Sami Kara from the Life Cycle Engineering and Management Research Group of the University of New South Wales for his fantastic ability to motivate and to give valuable guidance empowering me to formulate the right research questions. I would also like to acknowledge the academic contribution and supervision of Prof. Dr.-Ing. Prof. h.c. Klaus Dilger and Prof. Dr.-Ing. Michael Sinapius of the Technische Universität Braunschweig for enabling me to successfully finalise this book.

Special thanks goes to my colleagues Dirk Sauermann, Daniel Ebeling and Herwig Köster at the Volkswagen AG for providing invaluable insights into the complexity of enhancing energy efficiency in the automotive industry and for sharing their personal experience with me.

As a research associate, I have been blessed to work in teams with excellent colleagues, fellow researchers and friends eager to assist and support each other anytime. Thank you very much for the creative atmosphere and great teamwork experience. Among my colleagues, I am most grateful to Ms. Anne-Marie Schlake, M.A. and Dr.-Ing. Sebastian Thiede for their exceptional effort in reviewing my work and stimulating improvement in my *zeintific* approach. In addition, I want to express my appreciation to my research counterpart Mr. Wen Li, Ph.D., from the University of New South Wales for the excellent research cooperation and friendship that I am keen to maintain.

Above all, I thank my parents Birgit and Walter Zein for supporting me throughout my studies at university and most importantly providing me a home to work on this book. I cannot express how much I owe them for their vital encouragement and motivation enabling me to complete this work. Thank you!

Braunschweig, 19 June 2012 Dr.-Ing. André Zein

Contents

Abbreviations and Symbols

Abbreviations

Symbol	Value
Btu	British thermal unit
COLS	Corrected ordinary least squares
DEA	Data envelopment analysis
GDP	Gross domestic product
GPI	Green productivity index
MEEUP	Methodology study for Ecodesign of Energy-using Products
MRR	Material removal rate
OPC	OLE for process control
PLC	Programmable logic controller
SDEA	Stochastic data envelopment analysis
SFA	Stochastic frontier analysis

Indexes

Symbol	Unit	Value
E_{in}	Wh	Input energy demand
E_{loss}	Wh	Energy loss
E_{recover}	Wh	Recovered energy demand
P	W	Power consumption
P_c	W	Cutting power
P_C	W	Power of control unit
P_D	W	Power of dresser
$\mathrm{P}_{\text{fixed}}$	W	Fixed power consumption
P_{HS}	W	Power of hydraulic and cooling unit
P_S	W	Power of spindle
P_{SP}	W	Power of spindle pump
P_{total}	W	Absolute power demand
P_{WS}	W	Power of spindle and workpiece spindle
SEC	Wh/mm³	Specific energy demand

(continued)

Indexes (continued)

Symbol	Unit	Value
W_p	Wh	Process energy demand
W_f	Wh	Fixed energy demand
η	-	Machine tool efficiency
λ	mm^3/s	Material removal rate
λ_E	mm^3/s	Energy performance intensity metric
ν	mm^3/s	Material removal rate
ρ	mm^3	Removed material volume
σ	m/s	Cutting velocity

Symbols

Symbol	Value
W_{ijk}	Weighing factor
Z	Activity
P_E	Energy productivity
a_E	Energy consumption function
a_i	Consumption function
i	Consumption factor
k	Energy constant factor
n	Usage factor
r_E	Energy input
r_i	Input of a transformation
t_X	Waiting period
t_{RX}	Reactivation time
x	Output of a transformation
z	Technical capability

Chapter 1
Introduction

1.1 Problem Statement and Scope

Economic growth is bound to the consumption of energy (Berndt 1983; Metz et al. 2007). In the last three decades, each 1 % increase in global gross domestic product (GDP) was on average associated with a rise of 0.7 % in primary energy demand (International Energy Agency 2009). Forecasts of the U.S. Energy Information Administration anticipate the continuation of this trend with marginally decreasing intensity. In 2035, the primary energy demand is expected to rise by 0.56 % per marginal growth in GDP resulting in a total demand of 738.7 quadrillion Btu (U.S. Energy Information Administration 2010).

The transformation of primary energy from non-renewable resources as coal, oil and natural gas is inherently linked to carbon dioxide emissions. Carbon dioxide emitted from energy conversion represents the most abundant source of anthropogenic greenhouse gas affecting climate change. Therefore, energy-related CO_2-emissions are in the centre of environmental concerns (International Energy Agency 2009; European Commission 2008a; Metz et al. 2007).

With regard to the threat of environmental change and an impeding scarcity of non-renewable energy sources, the decoupling of economic growth and energy-related emissions has become imperative and intensified as new industrial economies emerge (World Commission on Environment and Development 1987; Meadows et al. 2001; Bruggink 2011). In response, over 75 countries have announced and begun to enact self-imposed limitations for greenhouse-gas emissions (United Nations 1998; International Energy Agency 2009). The European Union implemented for instance the regulatory framework *Energy 2020*, which ascertains an emission reduction target of at least 20 % by 2020, relative to 1990 levels. Long-term improvement is set to abandon 60–80 % of greenhouse gas emissions by 2050 (European Commission 2010). These targets have been reinforced by policies implementing an emission trading scheme, providing incentives for the expansion of renewable energy generation and defining targets for energy efficiency (European Parliament and Council 2009; European Commission 2008c).

A. Zein, *Transition Towards Energy Efficient Machine Tools*, Sustainable Production, Life Cycle Engineering and Management, DOI: 10.1007/978-3-642-32247-1_1,

The improvement of energy efficiency is a most cost-effective and immediate measure, also avoiding negative limitations for economic growth (Nolte and Oppel 2008; von Weizsäcker et al. 2010). To promote energy efficiency, ecodesign guidelines for energy-related products are implemented that define minimum electrical energy requirements and provide measures for electrical energy reduction. In industry, these guidelines are e.g. enforced for electric motors in order to exploit cost-effective energy-related improvements of 20–30 %. These potentials accumulate to 69 million tons of CO_2-emissions (European Commission 2009; German Electrical and Electronic Manufacturers' Association 2010). In the course of the recurring review process for the extension of the Ecodesign directive, machine tools have become subject to regulation due to substantial energy-related environmental impacts in the use phase and apparent potentials for improvement (European Commission 2008b). Estimates anticipate an electrical energy consumption of machine tools with up to 300 TWh/a in the EU-27, which is predominantly induced by 4.5 million operating machine tools for metalworking (Schischke et al. 2011b).

Machine tools are complex systems requiring energy to perform the geometric shaping of workpieces in a defined quality using appropriate tools and technologies (Deutsches Institut für Normung e.V. 2003; European Parliament and Council 2006; Tönshoff 1995). In machine tools, electrical energy is mainly converted into mechanical energy for operation. In order to improve the environmental performance of machine tools, energy-related measures have been developed and applied comprising organizational as well as technical aspects (Binding 1988; Devoldere et al. 2007). While organizational measures focus on the operating mode of components or the entire system, technical measures address the substitution of components through alternatives with improved power consumption (Zein et al. 2011; Wolfram 1986).

The structured consideration and integration of energy-related improvement measures in the design and for the operation of machine tools is yet absent (Schischke et al. 2011a). Best available technology is only partially identified and implemented (International Energy Agency 2006; Fruehan et al. 2000). Consequently, potentials remain unexploited even though the energy consumption of machine tools is also increasingly gaining economic relevance as part of the life cycle costs (Tam 2008; Fruehan et al. 2000; Schweiger 2009). The origins for the reluctance to improve the energy efficiency have initially been estimated in economic aspects, but must be extended towards information, communication and time barriers as well as deficits in competence (Nolte and Oppel 2008; Wanke and Trenz 2001). For machine tools, the exploitation of the improvement potential in energy efficiency fails due to lack of information on

- the machine- and process-specific energy consumption, which is determined by the configuration and operation of the machine system,
- the energy-related improvement potential for machine systems, which is quantified by comparing the actual demand with a minimum energy reference,
- the possible actions and their specific impact to improve energy efficiency.

1.2 Research Objective and Approach

Revising the deficiencies to advance energy efficiency, the central objective of this research can be formulated as follows:

The primary objective is the development of a performance management concept providing methods and tools to quantify and evaluate the energy efficiency of machine tools as a basis for the initiation of improvement towards an ideal energy level.

The proposed concept obtains a production theoretical scope as a methodological basis for determining minimum energy limits. The limits are derived empirically facilitating energy-related comparisons and rankings of machine tool systems. Furthermore, energy limits enable to quantify particular saving potentials and formulate improvement strategies. Consequently, this information encourages the application of improvement measures and promotes the development of energy efficient machine tools with minimum energy requirements.

The scientific objective of this research is to advance performance measurement as management discipline for the evaluation of energy efficiency and provide methodological guidance for the industrial application. This includes the integration of the concept as part of a continuous improvement process, which involves the machine tool builder being responsible for the machine design and the machine tool user regarding the operation of the system.

To achieve the objective, the development of the approach is structured in seven chapters. Following the problem statement and research objective, Chap. 2 introduces the context for improving the energy demand of machine tools. Initially, the origins of energy usage in machine systems are presented and linked with improvement measures highlighting the variability of measures and the barriers impeding the implementation. For the derivation of machine-specific improvement strategies, performance management is analysed as a method to assess the efficiency of transformation processes and pursue enhancement in energy demand.

The state of the research is reviewed in Chap. 3 giving specific attention to available approaches, which focus on the evaluation and realization of energy improvements for machine tools. For that reason, existing approaches are revised with regard to the ability to determine and assess improvement potential. By classifying and evaluating the approaches, limitations are addressed and preconditions for obtaining the research objective derived.

In the Chap. 4, the performance management concept for the energetic assessment and improvement of machine tools is introduced. Starting with the description of the concept layout, the concept structure is depicted including the characterization of the four concept modules in detail: determination of energy performance limits, measurement of actual performance, energy breakdown analysis and improvement planning. These elements are finally integrated as part of a continuous management process.

Chapter 5 describes the concept realization and prototypical implementation in an industrial context in order to assess the operability and verify the concept.

The outcome of the research is subject to review and critical evaluation in Chap. 6 before an outlook on concept extensions and recommendations for further research is presented. Chapter 7 concludes with a summary of the selected approach and research findings. The described research approach is illustrated in Fig. 1.1.

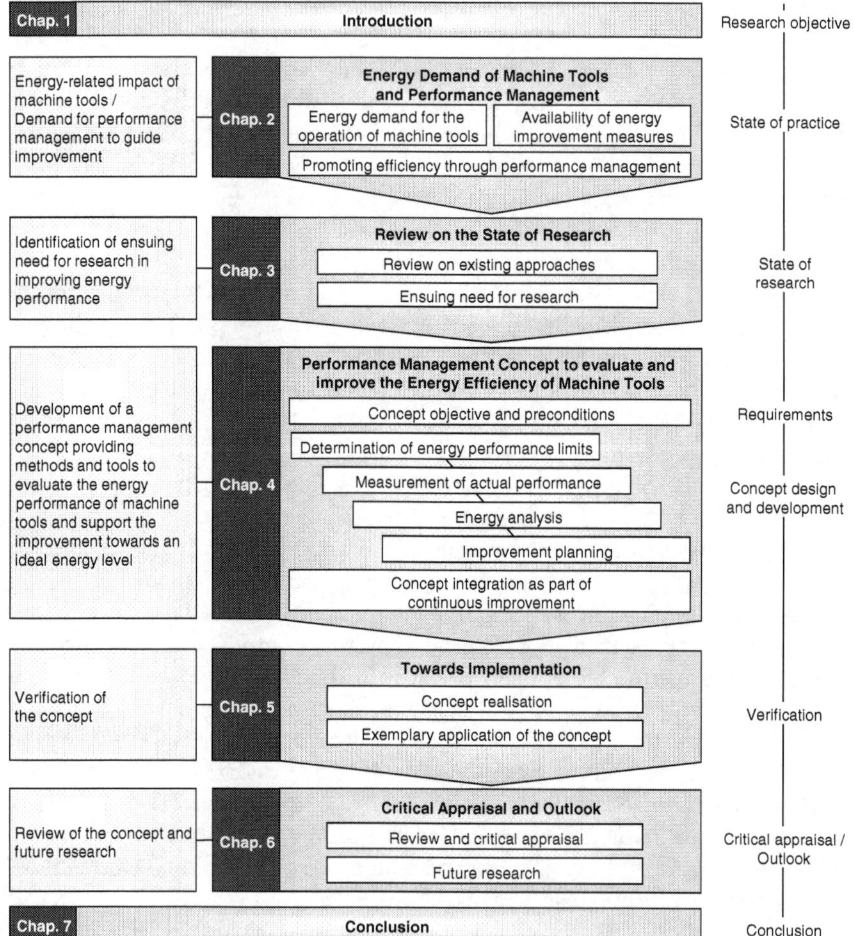

Fig. 1.1 Structure of the research approach

Chapter 2
Energy Demand of Machine Tools and Performance Management

This chapter introduces the thematic context as a basis for the elaboration of the research. First, the ecological implications of industrial growth in the economic system are presented, triggering the initiation of the concept of sustainable development. An analysis of energy flows in production systems is then given to identify the origins of energy demand and extended in detail for machine tools. Based on an assessment of energy improvement measures for machine tools according to the availability and barriers, the implications of imperfect information on improving the energy demand are reflected. As a method for promoting efficiency, performance management is introduced and analysed in terms of the feasibility to determine minimum energy requirements and to guide improvement.

2.1 Implications of Energy Usage in Industry

Economic systems interact with the ecologic system through the exchange of input and output flows. The ecologic system represents the cradle of natural resources for the value creation processes within the economic system and the sink for emissions and waste (Dyckhoff and Souren 2008; Meadows et al. 2001). Resources are generally divided into renewable and non-renewable feedstock. While non-renewable resources like coal, natural gas and mineral oil are limited in availability, renewable sources as organic material can provide a regenerative supply. The consumption of resources and the release of emissions and waste facilitate the growth of the economic system. Especially, energy resources are an indispensable input factor of the economic system for value creation (Singh et al. 1998; Nolte and Oppel 2008).

The world's primary energy demand was about 495 quadrillion Btu in 2007. The industrial sector accounts for more than 50 % of the energy used with more than 249 quadrillion Btu and dominates the commercial, residential and transportation sector. The demand for primary energy in industry is allocated to five energy sources, with electricity taking up the largest share of 37 % (see Fig. 2.1).

A. Zein, *Transition Towards Energy Efficient Machine Tools*, Sustainable Production, Life Cycle Engineering and Management, DOI: 10.1007/978-3-642-32247-1_2,

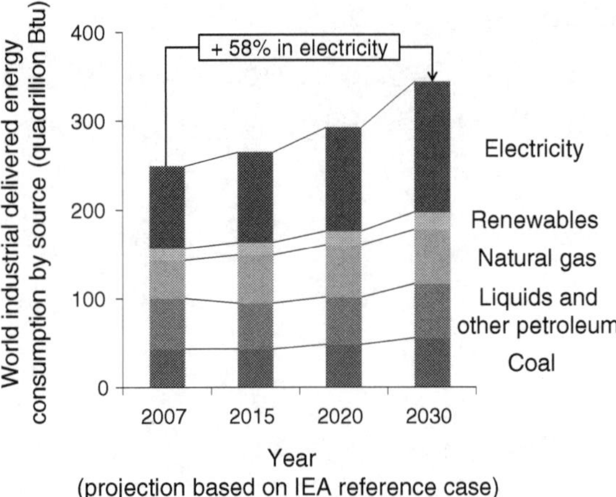

Fig. 2.1 Projection of world industrial energy consumption by sources, based on (U.S. Energy Information Administration 2010)

This includes the direct use of electricity as well as the losses for the generation, transmission and distribution. The industrial primary energy demand is expected to increase by 38 % to 344 quadrillion Btu until 2030, compared to the year 2007. Electricity remains the essential energy form in 2030 with 43 %, an increase of 58 % relative to 2007 (U.S. Energy Information Administration 2010).

Due to the strong reliance on fossil resources, the conversion of energy in the economic system is inherently linked to carbon dioxide emissions. In order to describe the impact on the ecological system, the industrial primary energy demand and the associated carbon dioxide emissions of economies with different economic performance are compared. The economic performance is expressed through the Gross Domestic Product (GDP) per capita, a common measure for the

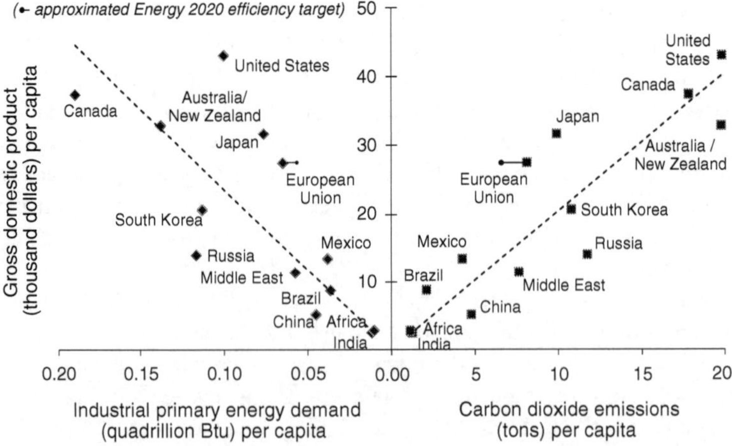

Fig. 2.2 Trends in industrial primary energy demand and energy-related carbon dioxide emissions, based on (U.S. Energy Information Administration 2010)

standard of living (Goossens et al. 2007). Figure 2.2 illustrates the GDP in relation to the primary energy demand and energy-related carbon dioxide emissions for 13 economies (referenced 2007) (U.S. Energy Information Administration 2010). The results of the analysis show that the industrial energy demand and emissions correlate in a linear fashion with the economic performance. Each 1 % increase in GDP is associated with an increase of primary energy by 0.05 % (the equivalent to 5 trillion Btu) and 0.54 % in carbon dioxide emissions.

With regard to the growth of (emerging) economies, the increase of the primary energy demand and the emissions of carbon dioxide as the most abundant greenhouse gas are expected to rise (International Energy Agency 2009). Studies of the *Club of Rome* revealed already in 1972 that the impact of economic growth goes beyond the capacity of the ecological system (Meadows et al. 2005). While the availability of fossil resources and the renewal rate of biogenic resources limit the use of primary energy in the economic system, the ecological system is bound by the ability to absorb waste and emissions without endangering the stability of the ecosystem (Meadows et al. 2005). The threat of climate change and the impending scarcity of energy sources led therefore to the elaboration of the Concept of Sustainable Development by the United Nations:

Sustainable development is development that meets the needs of the present without compromising the ability of future generations to meet their own needs. (World Commission on Environment and Development 1987).

The pursuit of a sustainable development has a broad scope taking into account economic, ecologic and social dimensions. The ecologic dimension is concerned with the interaction between the economic and ecologic system. It concentrates on the preservation of the ecosystem as well as the compliance of economic activities with the capacity of the ecologic system to provide resources and absorb emissions (Herrmann 2010; Dyckhoff and Souren 2008). Three strategies can generally be differentiated to enhance the ecological sustainability (Huber 2000; Dyckhoff and Souren 2008):

- *Sufficiency* refers to the self-determined limitation of all activities in the economic system in order to reduce the impact on the environment. It is implicitly noted in the definition of a sustainable development proposing the renunciation of consumption as part of an ecologically appropriate life style.
- *Efficiency* is inspired by the concept of technological progress and aims at the improvement of the input–output relations of existing transformation processes towards minimum input or maximized output levels. It is considered as the most cost-effective strategy providing immediate benefit.
- *Consistency* means the adaption of transformation processes ensuring coherence and compatibility of the renewable and non-renewable flows between the economic and ecologic system.

In order to envision thoroughly the ecologic dimension of a sustainable development, these strategies are specified through rules of conduct as well as legal directives (Dyckhoff and Souren 2008; International Energy Agency 2009). One of several examples is the regulatory framework *Europe 2020*, which specifies the three strategies of the ecologic dimension in terms of increasing energy efficiency, expanding the energy conversion from renewable sources and reducing greenhouse gas emissions by 20 % until 2020, relative to 1990 levels (European Commission 2010). To illustrate the implications of this framework on the European industrial sector, the mandatory reduction in the primary energy demand and carbon dioxide emissions are approximated for the increase in energy efficiency in Fig. 2.2 based on data from 2007 (U.S. Energy Information Administration 2010, 2007). Under the assumption of a steady GPD per capita until 2020, an increase in energy efficiency by 20 % demands a reduction of 13 % in the primary energy demand. Regarding the energy-related carbon dioxide emissions, this leads to a mandatory abatement of 19 % relative to 2007 in order to comply with the specified strategies.

To achieve the reduction in primary energy demand carbon dioxide emissions, a set of policies is extensively being developed and applied in order to amplify the adoption of energy-related instruments and the incorporation of energy efficient technology in the industrial sector. These include energy audits and energy management schemes as well as comparative assessments of the energy consumption for energy labelling (Verein Deutscher Ingenieure e.V. 1998; International Standard Organization 2011; European Parliament and Council 2009; European Commission 2010). The intention of these policies is to provide support for the implementation of (management) processes and systems in the industrial sector that help to improve the usage of energy (International Standard Organization 2011). An initial prerequisite for energy-related improvement described in all policies is to gain awareness about existing energy flows in order to identify improvement potentials (Herrmann 2010; Müller et al. 2009).

2.2 Energy Flows in Transformation Systems

Energy is a scalar quantity that exists in different forms as chemical, electrical, magnetic, mechanical or thermal energy. It is indirectly determined through measuring energy changes (Dincer and Rosen 2007). The different forms of energy can be converted via work and heat into one another at the expense of degradation in quality due to the thermodynamic irreversibility (Atkins and de Paula 2010). With respect to the conversion, energy can generally be distinguished into primary energy (e.g. as embodied energy in coal) and secondary energy (e.g. transformed into electrical energy) (Harvey 2010).

Energy is a mandatory input for industrial production. It can only partially be replaced by other resources (factor of peripheral substitution) (Rager 2008). Industrial production encompasses all activities that create value through the transformation of resources, components or parts to desired goods (Bellgran and Säfsten 2010; Dyckhoff and Spengler 2010). The terms *production* and *manufacturing* are often used interchangeably (Kalpakjian and Schmid 2001). To differentiate both terms, a systems-based distinction is chosen in accordance to the

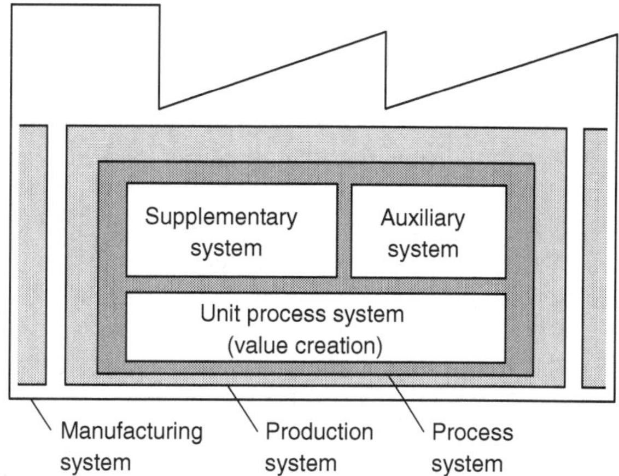

Fig. 2.3 Perspectives for energy analysis in manufacturing systems, adapted from (Bellgran and Säfsten 2010; Schieferdecker 2006)

definition provided by the International Academy for Production Engineering (CIRP).

Manufacturing is defined as … a series of interrelated activities and operations involving the design, materials selection, planning, production, quality assurance, management and marketing of the product of the manufacturing industry (CIRP 1990).

Production is … the act or process (or the connected series of acts or processes) of actually physically making a product from its material constituents, as distinct from designing the product, planning and controlling its production, assuring its quality (CIRP 1990).

Based on these definitions, a manufacturing system entails the activities within a plant or factory. It embraces the production system, which consists of the elements process, operand and operator. The operand is transformed in the process element, which is guided by the operator. The interrelation between these three elements determines the structure of the production system and affects the transformation process (Bellgran and Säfsten 2010). The process element can be further diversified into supplementary, unit process and auxiliary subsystems (see Fig. 2.3) (Schieferdecker 2006). Unit processes are decentralized operating entities, which directly use energy to perform the designated value creation. An indispensable precondition for the unit processes is the operation of supplementary systems. These are centrally operating entities that convert supplied energy in other forms meeting the requirements and demands of the unit processes. This includes systems that provide compressed air, filtering devices for cutting fluids or voltage transformers. The function of the auxiliary system is to maintain the operability of the entire system (e.g. ensure thermal and visual comfort) (Rager 2008). The supplementary systems, which are not directly associated to the unit processes, and the auxiliary systems are jointly considered as technical building services (Herrmann 2010).

From an energy-oriented perspective, these three entities of the process system represent the origins of energy demand in manufacturing systems, which accumulate on higher system levels to the absolute demand. To analyse the energy demand in production, two methodological approaches can be distinguished. While the energy-related assessment on production or manufacturing system level is referred to as macro-analysis operating with aggregated data, the objective of a micro-analysis is the detailed specification of the energy demands in unit process systems (Binding 1988).

On average, manufacturing systems demand chemical and electrical energy, which is transformed within the supplementary and auxiliary systems into thermal and mechanical energy (Müller et al. 2009). A review of the environmental statements of three Volkswagen automotive manufacturing systems exemplifies that electrical energy is the dominating form of energy with 48–65 % of the total demand (Volkswagen 2011a). While this aggregated data is commonly available, an assignment to the entities on the process system is absent. Therefore, a case study was performed for a metalworking production system in the automotive industry intending to disaggregate empirically the energy demand. The study included the unit process and supplementary systems with more than 15 machine tools as well as three centralized filtering devices and three mist collectors for cutting fluids (excluding the compressed air systems). The results point out that 56 % of the total electrical energy is used in unit processes and 44 % in supplementary entities. An alternative case study for a different metalworking production system in the automotive industry confirms the distribution of the energy shares (Bode 2007). Since a substantial share of electrical energy consumption originates in unit process systems, the electrical energy demand of these systems is subsequently analysed as a basis to deduce energy-related measures for improvement.

2.2.1 Electrical Energy Demand of Machine Tools

In production systems, inputs are transformed into tangible outputs through a sequence of technological processes. These processes are carried out by machinery. Machine tools represent a distinct class of metalworking machinery, which are defined as stationary operating, assembled systems fitted with a drive system other than directly applied human effort (Schischke et al. 2011a). They consist of joined parts and moving components enabling the entire system to perform a complex, useful function, which is the geometric forming, shaping or joining of workpieces in a defined quality using appropriate tools and technologies (European Parliament and Council 2006; Tönshoff 1995; Deutsches Institut für Normung e.V. 1985). The classification of machine tools relates to the structure and taxonomy for manufacturing processes according to the DIN standard 8580 (Deutsches Institut für Normung e.V. 2003; Brecher and Weck 2005).

A machine tool consists generally of a machine frame, guides, drives and control units (Tönshoff 1995). The integrated electrical components are signal elements, drives and actuators as well as wiring and measuring systems (Weck and Brecher 2006b). The electrical energy demand results from the temporal

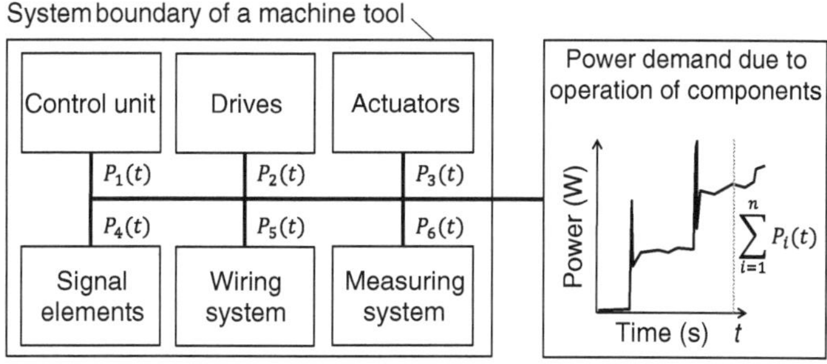

Fig. 2.4 Accumulation of power demands in a machine tool system (Zein et al. 2011)

accumulation of the individual power consumption for each component (see Fig. 2.4). Throughout the operation of the machine tool, process-induced performance requirements affect the power demand of the components. Depending on the structure of the machine elements and their operation, the power consumption is therefore not static but rather dynamic (Wolfram 1986; Bartz 1988).

The electrical energy demand of machine tools is rarely specified and known. Fractional information on energy requirements is provided for instance in life cycle inventory databases (e.g. Ecoinvent) for machining unit processes (Steiner and Frischknecht 2007). Yet, this information is restricted to a selection of machining processes and based on estimations providing aggregated and averaged data about the energy demand (Steiner and Frischknecht 2007). Figure 2.5 depicts the available inventory unit processes provided in the latest version of Ecoinvent supplied by the Ecoinvent Centre. With regard to the averaged values and underlying estimations in the existing inventory datasets, an in depth specification is generally recommended in case the scope centres the machining operation (Steiner and Frischknecht 2007).

In order to gain insight on the specific energy requirements of a machine tool, the energy demand is predominantly determined through power measurements. Power meters enable to capture the power demand by recording the current,

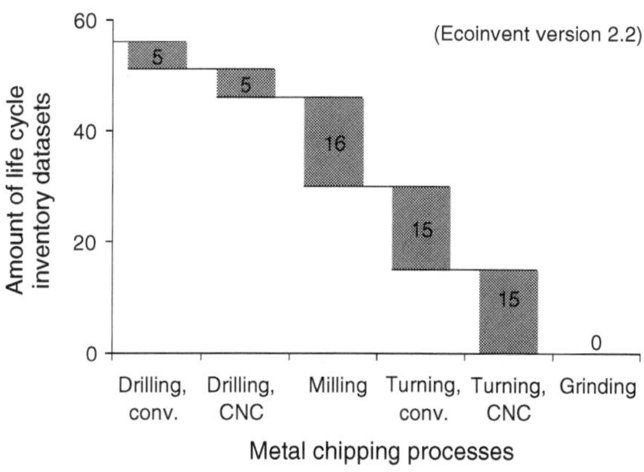

Fig. 2.5 Availability of life cycle inventory data, derived from (Steiner and Frischknecht 2007)

voltage and phase angle between voltage and current (in a three-phase system). By including the distortion in phase, the effective power is distinguished from the apparent power as the actual power delivered (Parthier 2010). The resulting power profile provides a basis to revise the operational modes and related power characteristics (Eckebrecht 2000). It is moreover an important instrument for time studies, which enable to distinguish productive and non-productive time shares of machine tool operation (Brecher et al. 2010; Kellens et al. 2011a).

Reviewing the power profile of an exemplary grinding process in Fig. 2.6, the activation of the machine tool and spindle as well as the material removal can be identified as operational modes. With regard to the power demand, a variable and fixed portion can generally be differentiated (Wolfram 1986; Dahmus and Gutowski 2004). The fixed power P_{fixed} covers the constant demand, which is necessary to ensure a functional mode of operation (e.g. waiting for operation). The process-induced portion relates to the power consumption for proceeding the machining process without touching the workpiece (so called air-cut) and the material removal capacity (Eckebrecht 2000; Wolfram 1986).

Power measurements represent the initial step to quantify the individual process energy W_p and fixed energy W_f, which are derived by integrating the power demands of machine tools over the processing period (1) and (2) (Binding 1988).

$$W_p = \int_{t_0}^{t_1} P(t) - P_{fixed}\, dt \tag{1}$$

$$W_f = \int_{t_0}^{t_1} P_{fixed}\, dt \tag{2}$$

A diversification of the fixed energy demand can be obtained from the power profiles by allocating power levels to the activation or operational modes of

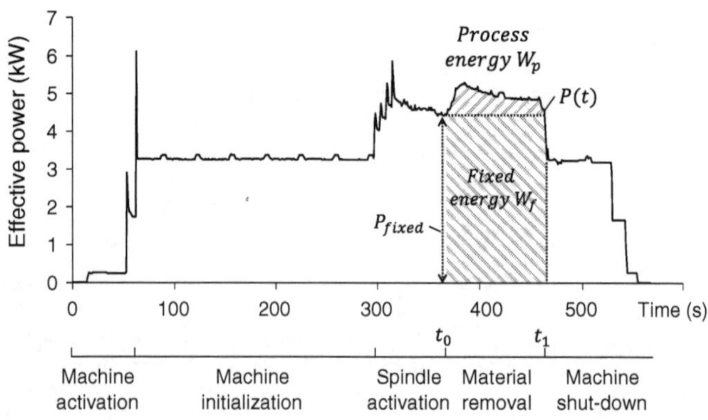

Fig. 2.6 Measured power profile for a grinding machine

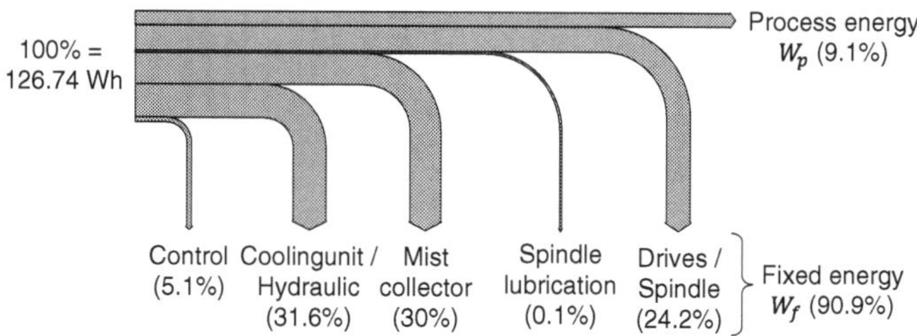

Fig. 2.7 Sankey diagram of energy flows for a material removal process

machine components (Eckebrecht 2000; Dietmair and Verl 2009a). In Fig. 2.7, the energy demands of components are visualized in a flowchart for the given grinding process based on the corresponding power characteristic during the machine activation, initialization and spindle activation. The allocation of energy demands to the components is exemplified in Appendix A.

As indicated in Fig. 2.7, the energy demand is determined by the process and individual machine characteristics. While the process energy depends on the material properties, selection of tools and process parameters ensuring an effective material removal, the machine characteristic is bound to the type of machine tool and the power demand of the integrated electrical components (Wolfram 1986; Binding 1988). In order to ease the effort to carry out power measurements, descriptive models have been developed that quantify energy demands based on material properties and technological parameters (e.g. chip thickness). These models rely on force prediction and enable to specify the tool tip energy for cutting the material (Rowe 2009; Kienzle 1952; Astakhow and Outeiro 2008; Wolfram 1986). This energy demand, however, takes up just a small share of the absolute energy. It is primarily used to analyse process restrictions and define technical requirements for the selection of spindles and machine tools (Kalpakjian and Schmid 2001). The extension of these cutting energy models towards predicting the process or total energy demand has been deficient in accuracy and applicability due to the machine-related variability in composition and specification (Wolfram 1986; Binding 1988; Dahmus and Gutowski 2004).

2.2.2 *Trends Affecting the Power Demand of Machine Tools*

Advances in manufacturing accuracy and processing performance have intensified the application of automation technology in machine tools. In addition to the automation of operational functions as handling or processing, modern machine tools have evolved to highly automated complex systems equipped with a variety of electrical means to monitor and maintain the operability and process quality (Weck and Brecher 2006a; Kalpakjian and Schmid 2001).

Power ratings represent as an attribute of machine tools a measure of the potential power use under individually assumed operating conditions. It is assigned by the machine tool manufacturer for an observed system designating for instance

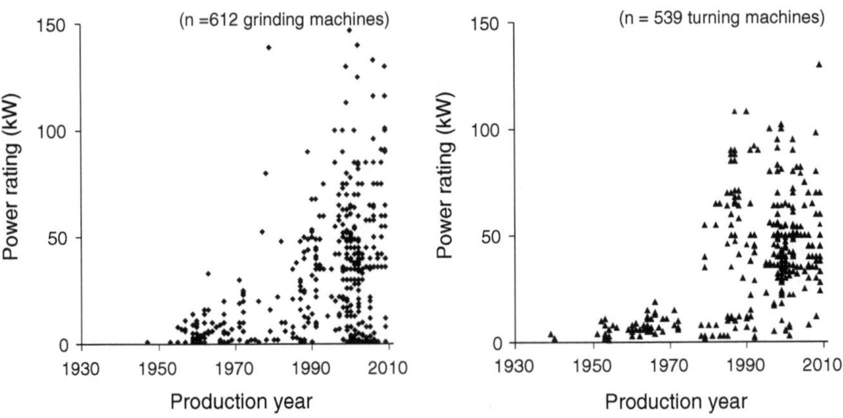

Fig. 2.8 Empirical long-term trends in power rating for machine tools

the minimum requirements for the power supply (Müller 2001). The power rating does not necessarily comply with the actual power demand providing nevertheless an indication about the power capacity (Li et al. 2011). Using the power rating as evaluative parameter, the effect of the increasing machine complexity and automation on the power consumption has been analysed for a total of 1,100 grinding and turning machines. The data is derived from three automotive metalworking manufacturing systems in Germany providing machine tool datasets for a time period of more than 70 years. Reviewing the long-term trend in Fig. 2.8, a substantial amplification in power ratings can for instance be determined for grinding and turning machines over the last three decades. The enveloping threshold in power rating increased from 24 kW around 1980 up to 100 kW with outliers even exceeding 147 kW. This trend appears to be interrelated to the introduction of computerized numerical controls in machine tools at the end of the 1970s (Degner 1986). It increased with the accelerated integration of electrical equipment (e.g. controls, sensors and actuators) improving the machining performance (Kalpakjian and Schmid 2001).

Besides the enhanced usage of automation technology, the origins of a higher power consumption of machine tools can be attributed to an increased power consumption of its integrated components. The demand for a higher processing performance has for instance been satisfied by increasing the torque of the main spindle resulting in higher power demands (Bode et al. 2008). As an example, the growth in output power of main spindles is illustrated in Fig. 2.9 indicating the trend for the last 25 years.

The results of these two empirical datasets specify a general upward trend in power consumption of machine tools induced by automation and component performance. Complementing the review of energy-related trends, an analysis of 30 machine tools performed by Degner provides an insight into the effective energy requirements to conduct the material removal (Degner 1986). The observations

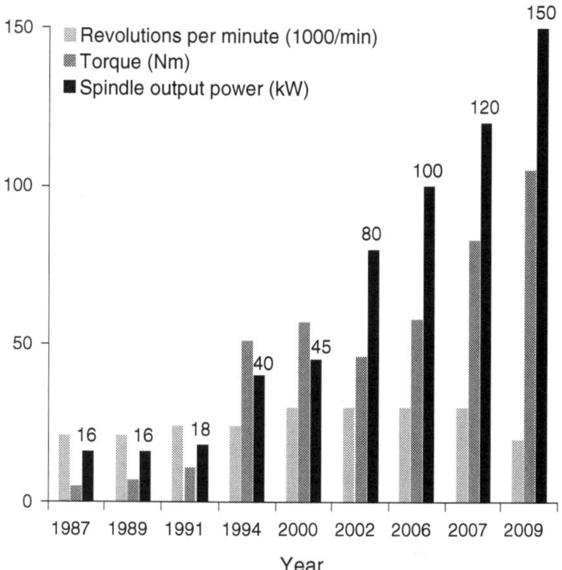

Fig. 2.9 Growth of main spindle performance (Reference Type HSK63) (Bode et al. 2008); updated in (Brecher and Bäumler 2011)

conclude for instance that the energy ratio of material removal to the absolute demand varies between 0.8–54 % with an average of 19 % among the observed machine tools. Degner deduces that the remaining share can accordingly be allocated to conversion losses within the components of the machine tool. In addition, the analysis points out the relevance of the fixed power exceeding the demand to conduct the material removal by a factor of up to 40 (Degner 1986). A comparison of two milling machines underlines this perception substantiating the increase in the fixed demand from a manual to an automated system (see Fig. 2.10) (Dahmus and Gutowski 2004). While the data provided by Dahmus and Gutowski and Degner reflects the power requirements of machine tools older than 20 years, it can be assumed that the correlations remain valid until today. This amplifies the relevance of the fixed power with respect to the described increase in power rating.

2.2.3 Improvement Measures for Electrical Energy Demand

With regard to the diversity of factors influencing the energy demand of a machine tool, a structured overview about available measures to improve the energy demand is obtained in this section. It is performed in order to provide insight about the obtainable potentials and focus point of measures. More than 100 energy-related measures have therefore been depicted from the Best Environmental Practice Manual as well as reports of the industry initiative Blue Competence and European Commission Product Group Study on Machine Tools (Schischke et al. 2012b; Prolima 2008).

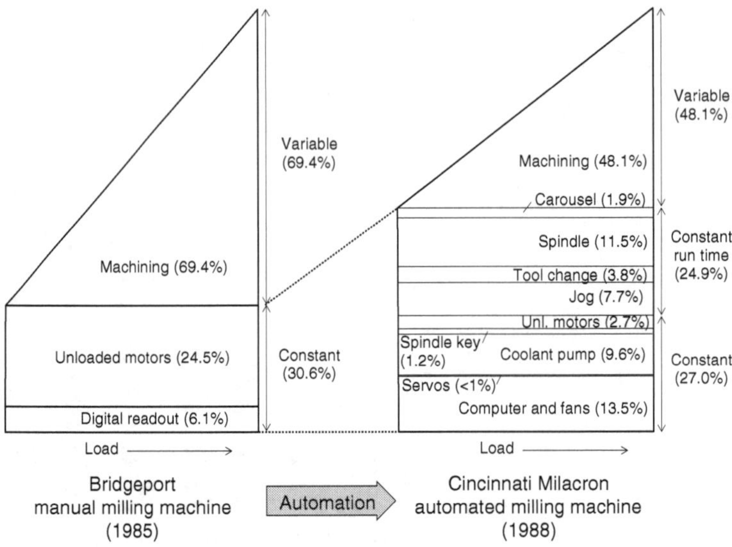

Fig. 2.10 Effect of automation on the fixed power demand (Dahmus and Gutowski 2004)

Improvement measures are fundamentally pursuing the objectives to minimize the energy losses in the provision, reduce the demand of the transformation and minimize the losses of the transformation (see Fig. 2.11). As power and time are the two factors directly relating to the electrical energy demand, improvements have either a temporal or a power-related characteristic. Measures to reduce the power consumption include for instance the replacement of machine components as hydraulics, drives or spindles with less power consuming alternatives (Neugebauer et al. 2008; Abele et al. 2011). Alternatively, the reduction of moving masses in machine tools using aluminium as lightweight material instead of steel has shown additional saving potential reducing the power demand in operation (Dietmair et al. 2010). In contrast, temporal advances aim at improving the

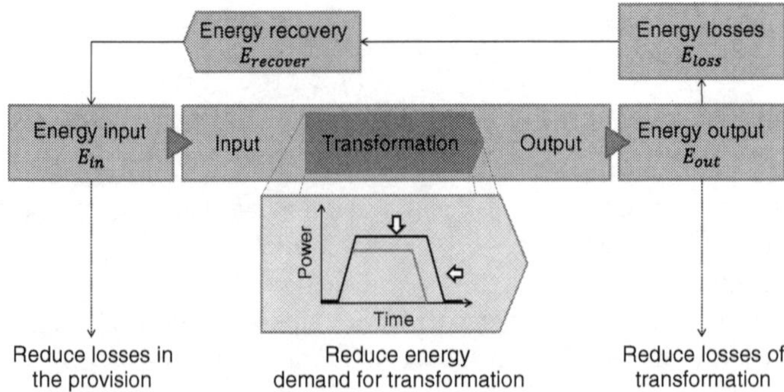

Fig. 2.11 Objectives of energy-related improvement measures, adapted from (Fanuc 2008; Müller et al. 2009)

machining task (e.g. near-net-shape machining) and the process design by increasing material removal rate or avoiding non-operational times in fixed power mode (Fanuc 2008; Heidenhain 2010; Müller et al. 2009). Synchronizing the acceleration and deceleration of a machining spindle and tool feed system to avoid idle times led for instance to an energy reduction of 10 % compared to the former operating conditions (Mori et al. 2011).

Energy losses E_{loss} represent the share of input energy E_{in} that is not contributing to the transformation and liberated as heat (Verein Deutscher Ingenieure e.V. 2003). Apart from the usage of heat emissions as a source for facility heating purposes, improvement measures predominantly intend to minimize the intensity of energy losses by reducing friction, leakages or process temperatures (Bartz 1988; Neugebauer et al. 2011). Additionally, measures can be implemented to recover energy $E_{recover}$ through kinetic buffering or heat recuperation (Müller et al. 2009; Diaz et al. 2010a).

Complementary to the consideration of objectives, measures can be allocated to the point of action comprising the process, component or entire machine tool. This structural assignment of measures enables to consider the effort and complexity to adopt an improvement (Binding 1988). While a process improvement can instantly be applied (e.g. variation of material removal rate), changes in the design of the machine tool or components are constrained and may therefore be contemplated in subsequent machine tool generations (Zein et al. 2011). Based on the energy-oriented classification of measures according to the objective and point of action, the identified improvement measures have been clustered. Figure 2.12 illustrates the selection of measures, which could clearly be assigned to the given set of categories. The results indicate a high availability of energy-reduction measures with a strong design emphasis of 40 % on the process, 35 % on the component and 25 % on the machine tool. Considering the reduction of energy losses due to

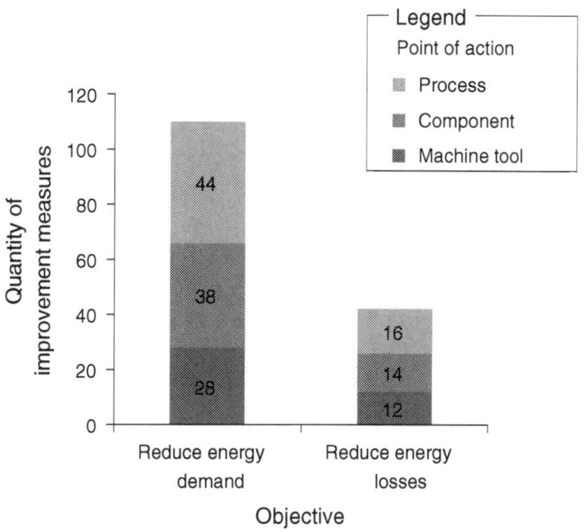

Fig. 2.12 Availability of improvement measures for machine tools, updated from (Zein et al. 2011)

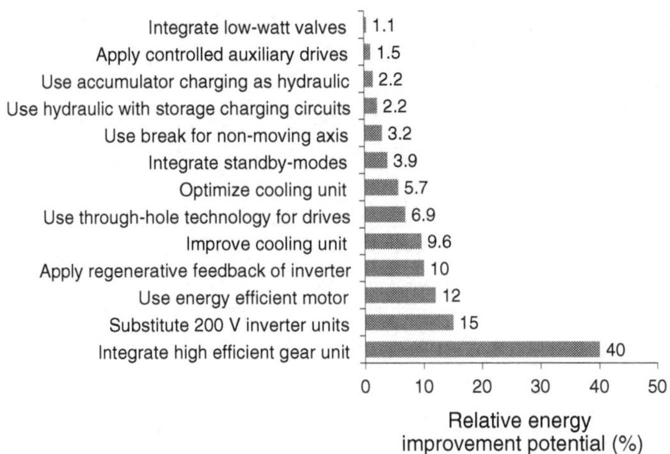

Fig. 2.13 Potentials of energy-related improvement measures, data derived from (Schischke et al. 2012b; Hegener 2010)

energy conversion, a minor share of measures was identified which can be attributed equally on all three points of action with limited variations.

Although more than 100 measures have been derived, information about the diffusion rate of measures in industry is restricted in availability. The same applies to information about the attainable energy savings, which is an essential prerequisite for the valuation of the profitability of improvement measures. An indicative ranking of measures has therefore been established providing an insight about the estimated relative improvement potential, which include savings of up to 40 % (see Fig. 2.13) (Schischke et al. 2012b; Hegener 2010). The absolute obtainable potential by aggregating a set of measures is sparsely exemplified for specific machine types. The results of prototypical implementations indicate an energy reduction of 30 % compared to the original machine system (Hegener 2010).

2.2.4 Energy Efficiency Gap

The motivation to enhance the energy demand is confronted with diverse barriers sustaining an "energy efficiency gap" (Jaffe 1994; Bunse et al. 2011). This phenomenon is defined as the deviation between the actual and ideal energy requirement of energy-using systems. It represents an unexploited improvement potential, which remains concealed due to failure and inertia to implement cost-effective measures (Baumgartner et al. 2006; Jaffe 1994).

Reviewing the literature on the origins of energy-related barriers inhibiting improvement, four aspects can generally be attributed regarding machine tools. These are specified as imperfect information, legal restrictions, profitability risks

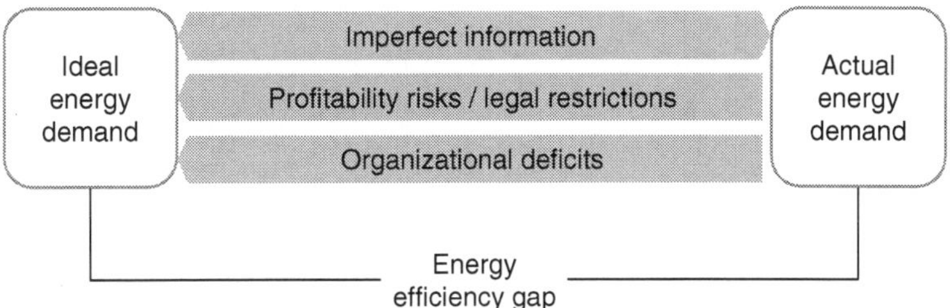

Fig. 2.14 Obstacles inducing the energy efficiency gap

and organizational deficits (see Fig. 2.14) (Nolte and Oppel 2008; Wanke and Trenz 2001; Baumgartner et al. 2006; International Energy Agency 2007; Seefeldt et al. 2006; Schleich 2009). The list is not exhaustive, but intends to provide an insight on the major causes contributing to the energy efficiency gap.

2.2.4.1 Imperfect Information

The attainment of energy-optimal operating conditions necessitates comprehensive information about the actual energy demand of machine tools and the benefits gained by associating adequate measures (Jaffe 1994).

Reasons for the lack of information about saving potentials originate in imperfect information about the current energy requirements and consumption patterns (Koopmans and te Velde 2001; Schleich 2009). This applies to information about the power demands as well as operating performance indicating the share of energy spend in productive and non-productive modes. The availability of energy-related information is commonly restricted due to the fragmentation and heterogeneity of machine tools and an unwillingness to perform time-consuming metering procedures (Wanke and Trenz 2001; Stasinopoulos et al. 2009).

The investment in energy improvements can furthermore fail due to imperfect information about suitable measures as well as limitations to evaluate the obtainable potential, reliability and applicability. The evaluation is usually aggravated by the prototypical implementation of measures in isolated case studies, which impedes the transfer and assessment of potentials in prevalent applications (Wanke and Trenz 2001). From this follows that as long as the performance of an activity and benefit of improvement measures remain unknown to the manufacturer or user, an implementation is unlikely to take place (Schleich 2009).

2.2.4.2 Legal Restrictions

The origins of legal restrictions are related to in the formalisms of business activities with third parties and legislative provisions. Restrictions can impede the implementation of energy improvement measures if warranty requirements and contract specifications are violated or additional testing and inspection regulations are implied (e.g. machine capability examinations) (Seefeldt et al. 2006).

2.2.4.3 Profitability Risks

From an economic perspective, failure to invest in cost-effective improvement measures can be substantiated by the inability to predict and quantify additional costs, which are associated with the implementation (Schleich 2009). These costs arise from all activities to obtain detailed energy-related information (e.g. metering power demands), to prepare and execute the measure as well as to revise the effect of the improvement. In total, these unpredictable costs can extensively exceed the envisaged energy savings (International Energy Agency 2007; Schleich 2009).

While up to 17 % of the life cycle costs can be associated for instance to the energy demand of a machine tool (see Fig. 2.15), the overall industrial energy costs take up however on average 2.2 % of the gross production value throughout the German metal processing sector (referenced to 2009) (Abele et al. 2009; Federal Statistical Office 2011).

This corresponds to 7.3 % of the labour costs and strengthens the perceived minor relevance of energy improvements obstructing investments (Wanke and Trenz 2001; Galitsky and Worrell 2008). Besides, alternative investments in quality or process technology are expected to provide higher benefits in compliance with restrictive payback periods than energy-related measures (International Energy Agency 2007; Seefeldt et al. 2006).

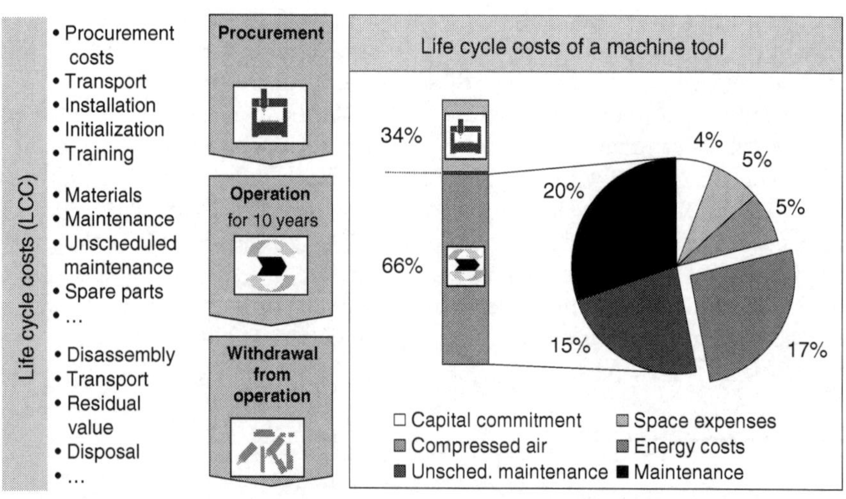

Fig. 2.15 Life cycle costs of machine tools, adapted from (Abele et al. 2009; Dervisopoulus et al. 2006)

2.2.4.4 Organizational Deficits

In addition to the given barriers, there are also organizational limitations relating to management constraints and split incentives. Lack of time, motivation and competence have been identified as the major management factors impeding the implementation of energy-related improvement measures (Nolte and Oppel 2008). Energy is therefore not among the strategic priorities of the top management indicating energy savings as rather incidental (Seefeldt et al. 2006; Schleich 2009).

Another organizational restraint which should not be underestimated relates to shared responsibilities impeding the overall incentive to invest in energy improvement measures (Wanke and Trenz 2001; International Energy Agency 2007). These limits arise from split incentives, when energy savings are not directly accountable to the originator and accrue elsewhere (Nolte and Oppel 2008; Schleich 2009). Split incentives prevail for instance between a machine tool builder and user about the integration of energy saving measures. It is only beneficial for the machine tool builder if additional profits can be gained. From a user perspective, investments may fail if information about the potential to recover the additional costs through energy savings is neither existent nor verified. Thus, imperfect information and the associated transaction costs impede to go beyond the organizational barrier (Schleich 2009; International Energy Agency 2007).

Despite the characterization of potential barriers obstructing the improvement of machine tools, the energy efficiency gap for machine tools has neither been evidenced nor quantified. This is mainly attributable to a lack of evaluative information about the ideal energy demand of machine tools as well as the actual energy demand of prevalent machine tools in operation. However, the variety of energy saving measures and associated potentials provide an indication that a considerable improvement potential remains unused, so far enabling only estimations about the magnitude of the energy efficiency gap.

2.2.5 Findings

Synthesizing the aspects that obstruct the energy-related improvement of machine tools, the importance to gain evaluative information about the energy efficiency gap is emphasized. The quantification of the energy efficiency gap is however bound to the ability of measuring the divergence of the actual energy usage in relation to the ideal energy reference providing insight on the associated magnitude of improvement potential.

Nevertheless, information constitutes solely a necessary but not sufficient condition for the realization of improvement. The sufficient condition comprises the operationalization of this information in management control mechanisms (DeCanio 1993). As a consequence, information and control represent the means to mind the barriers and guide the development of cost-effective improvement policies (Schleich 2009; Jaffe 1994).

2.3 Performance Management

Performance management strives to improve the performance of activities ensuring that actions are initiated to bridge a gap between the actual and designated ideal performance (Ferreira and Otley 2009). The evaluation of a performance gap through measurement constitutes the main objective of performance measurement as an integral element in the performance management process (Neely et al. 2005; Folan and Browne 2005).

Revising the fundamentals of the term performance, the process and the implementation of performance measurement as a means to gain evaluative information about performance is subsequently introduced in this section. Extending the scope towards the initiation of improvement actions, the integration of performance measurement into the process of performance management is described taking a cybernetics perspective on management control. By specifying performance management in an energy-oriented context, the capacity to determine ideal energy demands is reviewed as a basis to quantify and bridge the energy efficiency gap for machine tools.

2.3.1 Performance Measurement

Performance is an ambiguous term with diverse definitions adapting the specific context of applications and stakeholder perspectives (Hilgers 2008; Krause and Mertins 2006). Reviewing the commonalities among existing definitions, performance can be specified as the capacity of an activity to meet expectations. These are expressed by achieving a quantified target, which is evaluated in the dimensions effectiveness and efficiency (Erdmann 2002).

Effectiveness reflects the results of an activity as a qualitative measure. It obtains a strategic perspective ensuring that activities are initiated, which provide a contribution towards defined objectives (Hilgers 2008). In contrast, efficiency obtains an operational, transition-oriented perspective and valuates the degree of resource usage to meet an aspired result using quantitative metrics (Braz et al. 2011). Both measures combined form the evaluative dimensions of performance verifying the success of an activity (Lichiello and Turnock 1999).

In order to facilitate a better understanding of performance, the main characteristics have been deduced by (Krause and Mertins 2006). Accordingly, performance

- relates to relevant financial and non-financial characteristics and the gained benefit of a system,
- enables a multidimensional perspective,
- is influenced by the selected focus and thematic background,
- obtains a future- and action-oriented perspective on processes,
- originates in activities,
- and can be assessed using absolute or quantitative indicators (Krause and Mertins 2006).

The process of generating information about the effectiveness and efficiency of an activity is defined as performance measurement (Sturm 2000; Braz et al. 2011). According to Waggoner et al., it encompasses the

monitoring of performance, identification of areas that are in need of attention, enhancing motivation, improving communications, and strengthening accountability (Waggoner et al. 1999).

In addition to the evaluation of performance, performance measurement comprises also the establishment and implementation of processes to observe the performance and direct attention by providing information on a persistent basis (DeGroff et al. 2010; Stoop 1996). The process of performance measurement is described in the next section revising the elements and actions to formulate a target and to derive evaluative information about the performance of an activity.

2.3.2 Process of Performance Measurement

Reviewing the literature on the performance measurement process, an extensive variety of approaches can be identified. The elaboration of models for performance measurement ranges from brief descriptions on performed tasks, specifications of tools up to complete process guidelines (Bourne et al. 2003; Nudurupati et al. 2007).

A commonly proposed model to conduct performance measurement consists of four steps (Krause and Mertins 2006; Stoop 1996):

- *Diagnosis* encompasses all activities that relate to the review of objectives and requirements to implement the performance measurement. This includes also the definition of indicators, which are used to provide information about the performance in the subsequent steps.
- In the *projection* phase, the performance target is set as a quantitative indicator level expressing the commitment to achieve a defined performance.
- Within the *valuation* phase, the achieved performance is determined and compared with the performance target enabling to assess the effectiveness and efficiency of the observed activity.
- The final phase considers the *application* of the obtained performance information to pursue improvement.

In addition to this four step model, which is affiliated from the Deming-cycle, a three step procedure is anticipated by Stivers et al. and applied by Grüning. It sets a stronger emphasis on the measurement aspect within the performance measurement process (Stoop 1996). It stresses the *identification of performance factors, the measurement* and the *application* to pursue the defined target (Grüning 2002). Figure 2.16 depicts the commonalities between a comprehensive implementation approach in relation to the four and three step model. Having identified the three basic elements to conduct a performance measurement, the associated methods and tools performed in each step are subsequently described in detail.

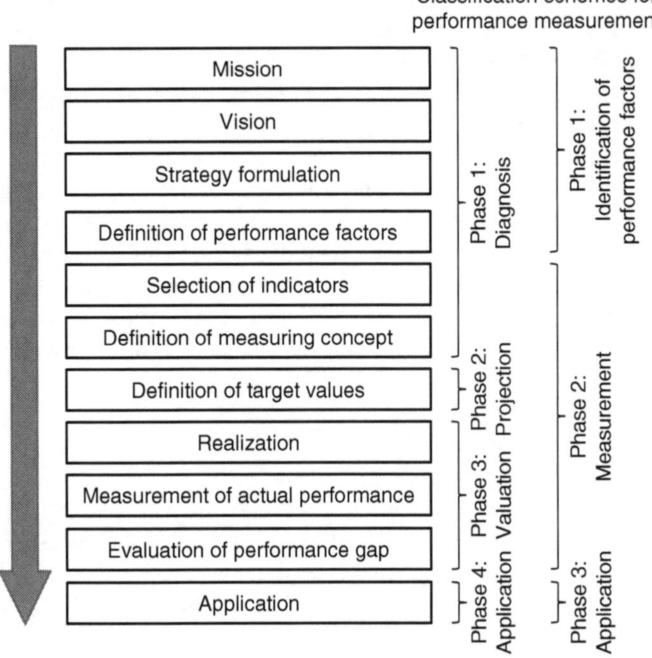

Fig. 2.16 Representation and classification of the performance measurement process (Grüning 2002)

2.3.2.1 Identification of Performance Factors

Measuring the performance of an activity demands initially to identify all aspects that affect the realization of an objective. These performance factors can for instance be extracted from strategies, which arise from the elaboration of visions and missions (Grüning 2002). From an economic perspective, five different types of performance factors are generally distinguished and categorized into traditional and modern factors. Physical and financial resources are jointly considered as traditional, well-established performance factors. In contrast, modern factors are addressing immaterial resources, processes and systems (Grüning 2002).

The selection of performance factors encompasses an evaluation of the relevance, importance and quality of the factor to represent the objective (Grüning 2002). In case multiple performance factors are identified, the interrelations between the factors have to be analysed in order to avoid ambiguous results from the performance measurement. By finally consolidating a structured set of performance factors, the objectives are operationalized into a performance measurement system, which builds the foundation for the subsequent measuring process (Hilgers 2008; Stoop 1996).

2.3.2.2 Measurement of Performance

The process of measuring performance starts with the conversion of identified performance factors into performance measures (Grüning 2002). Measures are metric indicators, which provide information about the characteristics of the performance factors in quantitative standards of measure (Blackburn 2008). Indicators can generally be expressed in form of relative numerical ratios and absolute terms (Jasch and Tukker 2009). An overview about possible indicators to measure the performance of processes is illustrated in Fig. 2.17. Accentuating a specific monitoring scope, it differentiates the indicators productivity, efficiency and profitability. The productivity metric obtains from a production theoretical perspective a purely physical input/output-perspective. The indicator on efficiency valuates the process by relating the expected resource usage to the actually used resources (Bellgran and Säfsten 2010). The profitability incorporates preferences regarding the economic implications (revenues and costs) of the transformation (Dyckhoff 2006). A thorough review on the interrelations between productivity and efficiency is provided in Sect. 4.3.2.

The selection of indicators has to ensure that the informational value about the performance factor is reliable and valid. The term reliability relates to the accuracy and reproducibility of the indicator to provide distinct information about the performance factor. The reliability of the indicator is a prerequisite for the assessment of validity. The validity of an indicator valuates the responsiveness and precision to react effectively to changes (Grüning 2002).

In addition to the informational value, the costs to determine and quantify the indicators are exerting a strong influence on the selection process. The measurement costs comprise all activities, which are necessary to collect the relevant data in order to appraise the indicator. Generally, the quantification of indicators for traditional performance factors is less expensive than for modern factors, as

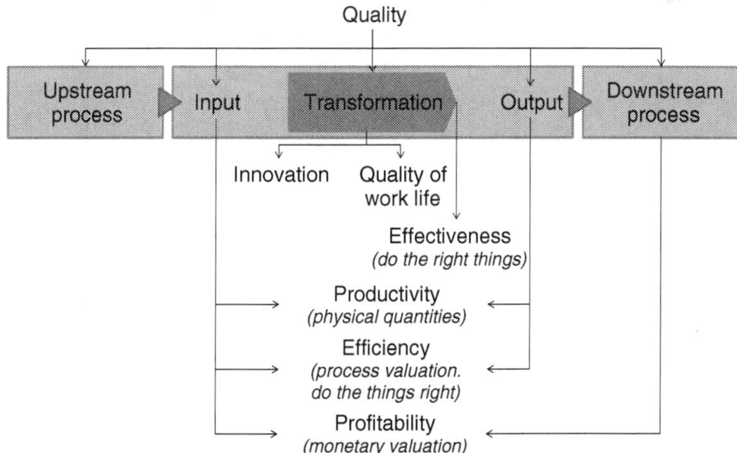

Fig. 2.17 Common indicators to measure the performance of processes, adapted from (Sink and Tuttle 1995; Bellgran and Säfsten 2010)

most indicators depend on existing data. Considering the selection of indicators, the reliability, validity and measurement costs have to be in alignment with the requirements to perform the measurement (Grüning 2002).

Following the determination of indicators, the process of defining target values and assessing the performance is conducted. By specifying a threshold value for each indicator, the performance targets are quantified representing a commitment of the anticipated level of performance (Stoop 1996). In order to ensure a successful operation of the performance measurement, it is essential to place an objective, realistic and challenging target value (Stoop 1996). Particular approaches to define and identify target values are subsequently introduced:

- Deriving performance targets from *intuition* is a simple method demanding little effort and costs. In situations in which no information about the performance is available, intuition enables to anticipate subjectively a threshold for the performance. However, this approach does not meet any of the given evaluative performance criteria. It should therefore only be applied once no other way of determining a threshold can be obtained (Stoop 1996).
- The introduction of a *monitoring* system supports the quantification of target values by reflecting former performance tendencies. This enables to determine an ideal target value based on observed performance levels. Implementing a monitoring system demands to consider that the performance threshold is concerned with the evaluation of the past. It does not enable to presume advances in performance due to improvement (Grüning 2002).
- *Quantitative models* inherit the characteristics of the evaluated system enabling to predict the performance. This allows an assessment of the actual performance under varying conditions and is the basis to derive the maximum attainable performance level. The usefulness of quantitative models depends however on the quality and the applicability of the model to perform accurate predictions (Nudurupati et al. 2007; Armstrong and Shapiro 1974). The evaluation of quality relates to the comprehensiveness of the underlying assumptions linking the real world and model. This also includes verifying the reliability of the predictions. Additionally, the applicability relates to the perceived quality of the model application indicated by the ability to resolve unknown information and the obtained benefit (Armstrong and Shapiro 1974; Stoop 1996).
- In addition to the quantitative models, target values can furthermore be obtained in form of *cooperative agreements* between all stakeholders, which exert influence on the performance (Grüning 2002). In this case, target values constitute essentially the willingness of stakeholders to contribute towards the improvement of performance (Bourne et al. 2003).
- *Strategic targets* represent another valuable source of information, which can be used to adopt performance thresholds. Nevertheless, it has to be noted that the conversion of strategic targets into performance targets contemplates a long-term perspective, which may require converting the targets for short-term periods (Grüning 2002).
- A final option to determine targets is the *comparative assessment.* It provides insight about obtainable levels of performance based on the comparison of

alternative solutions (Binding 1988). This enables on the one side to determine a common, averaged performance level, which sets the basis to define target values incrementally exceeding the quantified performance. On the other side, the comparative assessment strives to identify outstanding solutions among the revised alternatives, which represent best available solutions. The process to identify a reference with outstanding properties for comparative assessments is considered as benchmarking (Zhu 2009). The identified benchmark sets the target value that has to be obtained through performance improvements. An essential prerequisite to conduct a comparative assessment is in both cases the availability of alternative solutions, which are adequate for comparison (Stoop 1996).

Revising the capacity to determine objective, realistic and challenging target values, substantial differences and inabilities among all approaches become apparent. The capability of the methods to meet the selection criteria is approximated in Table 2.1 concluding the critical examination in literature. The definition of a target value is consequently considered as the most critical and challenging element in performance measurement (Centre for Business Performance 2004; Boyd et al. 2008). For that reason, the target value originates predominantly from empirical evidence in order to ensure the achievability in practical applications (Johnston et al. 2001).

Based on the specification of a target value, the actual performance is obtained through measurement. This demands the development of a measurement concept, which defines the procedures and tools to track the parameter value consistently. The actual performance is indicated by the obtained level of the parameter value on the indicator scaling (Grüning 2002). When considered individually, the actual performance level does not provide any use as a single indicator. Only through comparison with the corresponding target value, qualitative information on the performance is obtained (Cantner et al. 2007).

The evaluation based on the measured performance and defined target value is the starting point for the improvement process in the final step. It contemplates the

Table 2.1 Review of approaches to derive target values

		Approaches to define target values						
		Intuition	Monitoring	Quantitative models	Cooperative agreements	Strategic targets	Comparative assessment (average value)	Comparative assessment (benchmarking)
Properties	Objective	○	●	●	○	○	●	●
	Realistic	○	●	●	◐	◐	●	●
	Challenging	○	◐	●	◐	◐	◐	●

Characteristic value:

○ low ◐ average ● high

identification and implementation of activities to improve the performance striving to meet the target value.

2.3.2.3 Application of Performance Measurement

The consideration of performance measurement information in the context of decision-making is the last element in the performance measurement process. The results of the performance measurement provide evaluative information about prevalent gaps between the ideal and actual performance, which is used to derive and implement actions for improvement.

The use of information facilitates the valuation and communication of preceding performance trends serving the function to reward and motivate an impending, purposive behaviour (Krause and Mertins 2006). In addition, the results of the performance measurement enable a reconsideration of the feasibility to achieve the defined targets under the prevailing conditions. This can consequently lead to an adaption of target values (Hon 2005). The functional use of performance measurement is outlined in Fig. 2.18.

Providing information about the current performance and the obtainable improvement potential, performance measurement shares the characteristics of management control as element of the management process (Grüning 2002; Stoop 1996). The interrelation between the measurement of performance and the realization of improvement through management processes is therefore extended in the following section with the introduction of performance management.

2.3.3 Integration of Performance Measurement into Management

The measurement of performance constitutes improvement potential by quantifying the performance gap between an ideal and actual performance. The attainment

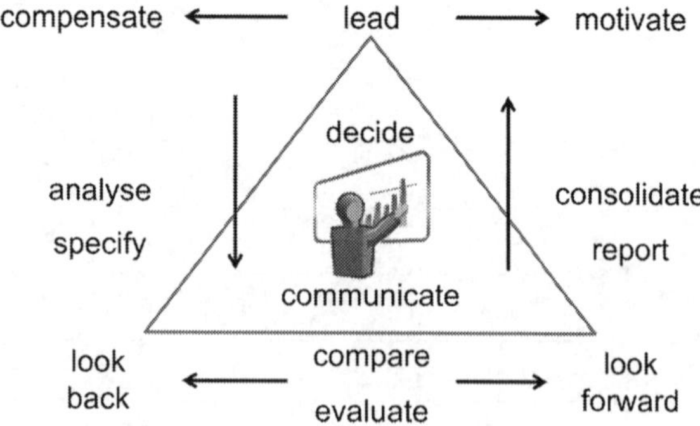

Fig. 2.18 Functions of performance measurement (Hon 2005)

of improvement to close the gap is in the responsibility of management (Lundberg et al. 2009; Folan and Browne 2005). Management can be defined as a process.

consisting of planning, organising, actuating and controlling, performed to determine and accomplish the objectives by the use of people and resources. (Terry 1968)

Accordingly, the management process encompasses four activities to initiate and adjust actions in order to realize a designated target (Mellerowicz 1963):

- Within the *planning phase*, methods and tools to achieve the target value are selected.
- The main task of the *organising phase* is to coordinate the resources ensuring the functioning of actions. This phase is combined in literature with the planning phase and aggregated as a joint phase (Grüning 2002; Günther 1991; Hilgers 2008).
- After the planning process has been completed, the implementation of actions is carried out constituting the actual progress towards achieving the target. This phase is therefore referred to as *actuating* or *directing*.
- In the final *controlling phase*, the outcome of actions is reviewed initiating control mechanisms, which steer the actions towards the designated target.

Taking a cybernetic perspective on management, the management process can fundamentally be understood as a control process, which strives to resolve problems in order to achieve and sustain an aspired, stable state (Gomez et al. 1975). The "science of control" is centred according to Wiener in the term *cybernetics* (Wiener 2000; Beer 1966). It belongs as a discipline to the theory of dynamic systems, which is concerned with the development of models based on systems as a conjunction of connected parts including their structures, interrelations and characteristics (Boulding 1956; Malik 2011).

The focus of cybernetics is on the control of systems and communication of information as a means to support the function of control (Otten and Debons 1970). The mechanisms enabling the control of systems are feedback and feedforward (see Fig. 2.19). Both inherit control strategies striving to abate the deviation from the desired target value enabling purposive actions and decision-making (Grüning 2002; Beer 1966; Mock 1986).

Feedback facilitates to reflect influences on the actual system state and to compensate the impact on the system in form of a negative information loop. In contrast, the concept of feedforward predisposes disturbances before they occur limiting the impact on the controlled system. The concepts jointly pursue the aim to support continuously the management process by adapting the input of the system (Mock 1986). Based on this perception, management is therefore considered as the "profession of control" (Beer 1966) and envisioned in form of a control loop.

In order to ensure an effective management control in practical applications, the adoption of the cybernetics approach demands information regarding the feedback and feedforward mechanisms affecting the dynamics and behaviour of the controlled system (Mock 1986). As a consequence, the central challenge to cybernetic

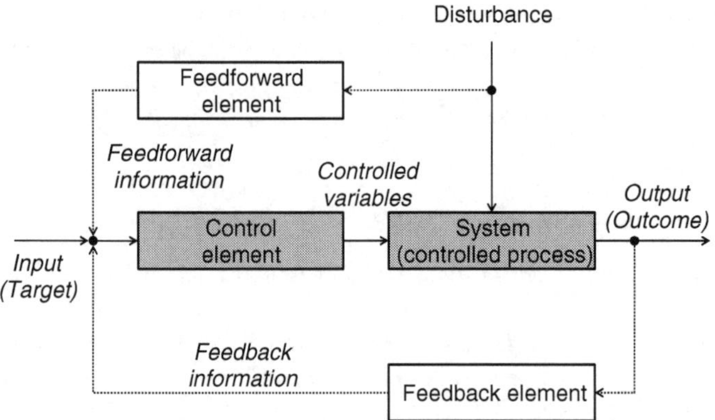

Fig. 2.19 Cybernetic control loops of management, adapted from (Grüning 2002; Baetge 1974)

management is to gain information in order to identify prevalent and evolving control mechanisms, which enable the attainment of the target value (Mock 1986).

Performance measurement can resolve this demand based on the results of the performance evaluation (Hilgers 2008). These enable to describe the effect of actions on the performance and therefore represent an important source of information, which can be used as input for the design of feedback and feedforward mechanisms (Hatry and Wholey 2006; Grüning 2002). Accordingly, performance measurement can be characterized as a means to specify management control as element of the cybernetic management process (Grüning 2002).

By actively using performance information within the control mechanisms, performance measurement is going beyond the evaluation of effectiveness and efficiency transitioning from the mere measurement towards management (Sturm 2000; Hilgers 2008). In accordance with the definition provided by Amaratunga and Baldry, performance management can be specified as

the use of performance measurement information to help set agreed-upon performance goals, allocate and prioritize resources, inform managers to either confirm or change current policy or program directions to meet those goals, and report on the success in meeting those goals. (National Performance Review 1997).

Given this definition, it is important to consider the dependence of performance management on performance measurement in an integrated context indicating the need for a performance management system (Folan and Browne 2005). Performance management systems comprehend methods and procedures in form of a framework ensuring the provision of relevant performance information to guide the management process towards the performance objectives (Krause and Mertins 2006). The common elements to plan, improve, measure and communicate performance as well as their interactions within a performance management system are illustrated in Fig. 2.20 showing the interrelations between the measurement and implementation activities.

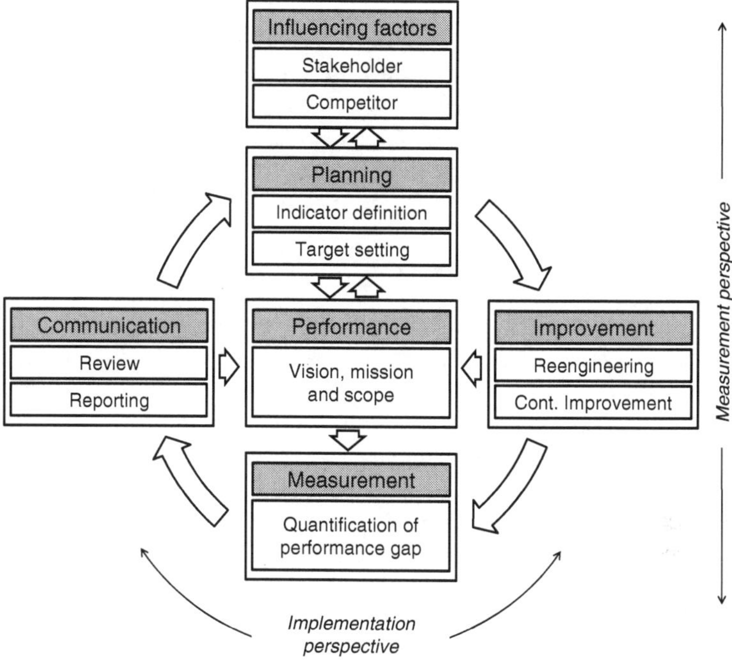

Fig. 2.20 Elements of a performance management system, adapted from (Krause and Mertins 2006; Hilgers 2008)

The illustrated activities of planning, improvement, measurement and communication form a self-learning process enabling progress through continuous improvement (Grüning 2002). Performance measurement and management can be seen in this management system as individual dimensions (Hilgers 2008). Both are connected to one another through the quantification of the performance gap. It embeds accordingly performance measurement as an element of management supporting the achievement of an initial target (Erdmann 2002).

2.3.4 Energy-Oriented Performance Management

The implementation and dissemination of performance management is advanced through a variety of particular economically and environmentally-oriented approaches representing individual frameworks to track the execution and pursue the improvement of activities (Folan and Browne 2005; Bourne et al. 2003; Neely et al. 2005). The approaches predominantly encompass monetary performance measures for strategic management levels as well as non-monetary measures on operational levels (Erdmann 2002; Dhavale 1996). The purpose of this section is to revise specifications of performance management systems for the energy usage in transformation processes providing methods and tools to quantify and bridge the energy efficiency gap of machine tools. This includes especially to differentiate the ambiguous term energy performance as well as to review means for the derivation of energy-oriented targets for performance measurement.

2.3.4.1 Energy Management System

Guiding the improvement of energy performance is a fundamental element of the recently published ISO 50001 on Energy Management Systems (International Standard Organization 2011). Originating from the ISO standards 9001 and 14001, the ISO 50001 bears a strong resemblance with the established management system standards. It can conjointly be integrated by organisations as part of an overall management system (International Standard Organization 2011). It defines a standardized framework of systems and processes enabling organisations to manage and enhance the energy performance (see Fig. 2.21).

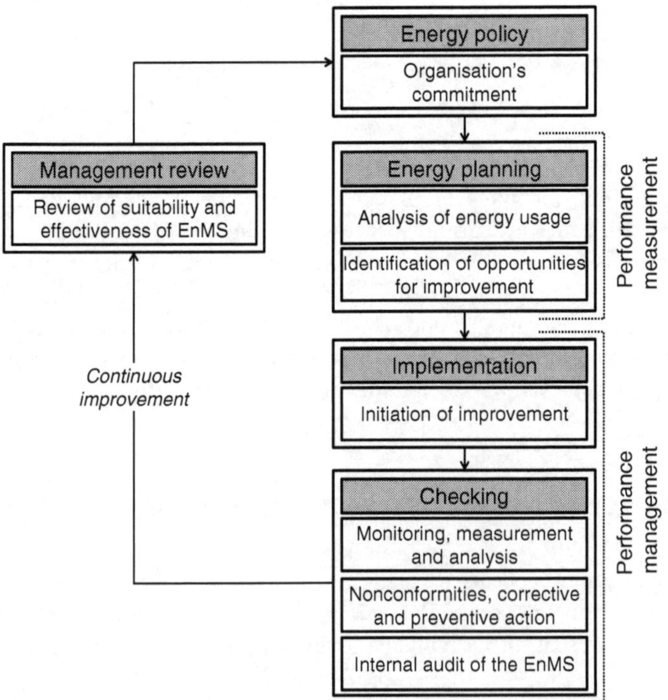

Fig. 2.21 Energy management system model, adapted from (International Standard Organization 2011)

Incorporating the elements of a performance management system, an energy management system (EnMS) encompasses the activities (International Standard Organization 2011):

- to implement an energy policy defining the organization's commitment,
- to perform an energy planning with the aim to identify and specify improvement potential,
- to initiate and implement improvement actions,
- to check the measurement and improvement of energy performance
- as well as to review the suitability and efficacy of the EnMS striving for continuous improvement of the management system.

With the definition of the mission, vision and scope of the EnMS, the energy policy corresponds to the initial element of the performance management system. Additionally, the energy planning complies with the activities of the performance measurement enfolding both the planning and measurement elements (see Fig. 2.20). It provides subsequently decisive information initiating the performance management in the implementation and checking activities.

By formulating the objectives and energy targets, the energy planning delineates the intention of the organisation and aspired commitment for improvement in energy performance. A prerequisite for transposing the objectives into measurable energy targets is however the specification of energy performance. According to the ISO 50001, energy performance comprehends "*measurable results related to energy efficiency, energy use and energy consumption*" (International Standard Organization 2011). It is consequently reflected through energy performance indicators representing quantifiable metrics (International Standard Organization 2011).

Energy efficiency is defined as the "*ratio or other quantitative relationship between an output of performance, service, goods or energy, and an input of energy*" (International Standard Organization 2011). Evading a distinct definition of the term, the ISO standard enumerates in the further course divergent examples as surrogates without any declaration of the underlying metrics. The lack of an unequivocal definition is also evident for the absolute energy performance indicators. It describes *energy use* as a "*kind of application of energy*" (International Standard Organization 2011) and *energy consumption* as the "quantity of energy applied" (International Standard Organization 2011).

With regard to the definition of an energy target as a measurable threshold for energy performance, the ISO standard refers to benchmarking as a methodological approach to gain quantitative input. In order to derive evaluative information about the achieved progress in energy performance, the EnMS proposes alternatively to compare the energy performance indicators with an energy baseline. Indicating a quantitative reference value of energy performance, the energy baseline is derived and updated through monitoring of energy demands in time studies (International Standard Organization 2011).

The preceding illustrates that the ISO 50001 uses main elements of performance management. It provides a general framework guiding the implementation of activities to pursue improvement in energy performance. Providing the basic structural elements of a management system, it does however not define distinct requirements for energy performance indicators and formalize methodological approaches for the design of energy performance targets (International Standard Organization 2011). An essential requirement for the implementation of an energy-oriented performance management remains therefore to decide on suitable indicators for the characterization of energy performance and means to derive energy targets.

2.3.4.2 Specification of Energy Performance Indicators and Energy Targets

Energy efficiency is an ambiguous energy performance indicator with a particular diversity of interpretations, which differ both in structure and in contextual form (Patterson 1996; Diekmann et al. 1999). Extending the examples provided in the ISO standard, a selection of applied energy efficiency indicators is compiled by Bunse et al. and Diekmann et al. in order to emphasize the wide scope of technical and economic variations (Bunse et al. 2010; Diekmann et al. 1999). An excerpt of identified metrics is comprised in Table 2.2.

Unifying characteristics among the ratios enable to differentiate classes of energy efficiency indicators in order to gain consistency among the various expressions (Diekmann et al. 1999). Patterson structures a first class of indicators for energy efficiency in form of dimensionless quantitative ratios (Patterson 1996). These can generally be subdivided into three segments reflecting the specific context of energy efficiency (see Table 2.3) (Patterson 1996; International Atomic Energy Agency et al. 2005). In line with the principles of thermodynamics, the first two segments designate ratios referring to the energy quantity and the energy quality. Defining the energy output relative to the total energy input, the scope of analysis for energy quantity concentrates on the transfer of energy in technical systems (e.g. electric motors). In contrast, the estimation of energy efficiency with

Table 2.2 Excerpt of energy efficiency indicators, based on (Bunse et al. 2010; Diekmann et al.1999; Patterson1996; Neugebauer et al. 2010)

Indicator	Characterization
Conversion efficiency	Energy output per energy input
Degree of efficiency	Net energy per used primary energy
Energy consumption intensity	Energy consumption per physical output value
Energy efficiency	Useful output per energy input
Energy efficiency factor	Process energy demand per machine energy demand
Energy intensity	Energy demand per unit of industrial output
Energy productivity	Output quantity per energy input
Final energy efficiency	Energy savings by the same benefits
Specific energy consumption	Energy demand per tonne of product material
Specific power consumption	Power consumption at a working point (or avg. of a cycle)
Thermal energy efficiency	Energy value available for process per unit energy value

Table 2.3 Classification of performance indicators for energy efficiency, adapted from (Patterson 1996)

Classes of energy efficiency indicators	Segments	Ratio	Units
Dimensionless quantitative rations	Thermodynamic (Energy quantity)	$\frac{\text{Useful energy output}}{\text{Total energy input}}$	[-]
	Thermodynamic (Energy quantity)	$\frac{\text{Actual energy input}}{\text{Ideal energy input}}$	[-]
	Economic	$\frac{\text{Economic quantity value}}{\text{Energy input costs}}$	[-]
Energy intensity (EI)/ specific energy consumption (SEC)	Physical-Thermodynamic	$\frac{\text{Energy input}}{\text{Physical quantity value}}$	e.g. [J/g]
	Economic-Thermodynamic	$\frac{\text{Energy input}}{\text{Economic quantity value}}$	e.g. [J/€]

regard to an ideal energy limit is concerned to assess the quality of energy conversion. A third segment of energy efficiency indicators encompasses monetary valued ratios, which are primarily used to illustrate aggregated data about economic activities (Diekmann et al. 1999).

Despite the dimensionless indicators, the energy intensity (also defined as specific energy consumption) describes a second class of performance indicators for energy efficiency (see Table 2.3). Relating the energy demand to a numerical quantity value, the energy intensity emphasizes explicitly the usage of energy to conduct a transformation (Diekmann et al. 1999). Generally, physical and economic characterizations are observed to define energy intensity.

Revising the composition of the indicators, it becomes apparent that only the ideal energy demand of the energy quality indicator instantly provides an evaluative reference for energy efficiency. In contrast, all alternative performance indicators rely on the comparison with defined reference thresholds or the monitoring of relative changes to evaluate progress and the achievement of targets (Patterson 1996; Erlach and Westkämper 2009).

As a means to derive thresholds for energy targets, the Methodology study for Ecodesign of Energy-using Products (MEEUP) designates an alternative approach to evaluate the performance of energy-using products through comparative assessment (Tanaka 2008; Kemna et al. 2005). This methodology is applied in the context of the Ecodesign directive initiating energy performance regulations to improve the environmental impact within energy-related product categories (Kemna et al. 2005; European Parliament and Council 2009). To conduct a comparative assessment according to the MEEUP, the development of an individual evaluation procedure is required for each product category under study. The empirical comparison relies additionally on market available products requiring therefore to select representative entities within the evaluated product category. Defining particular test cycles and real-life operating conditions, the energy demands within the product group are assessed enabling the identification of best available solutions for the given evaluation procedure (Kemna et al. 2005). This approach has for instance been successfully adopted for electric motors (de Almeida et al. 2008).

The ideal energy demand of the thermodynamic indicator and the comparative assessment represent in conclusion two approaches to define unbiased energy targets for energy performance management. While the thermodynamic indicator defines a theoretical, unattainable energy target, the comparative assessment enables to cope with this limitation by finding best available solutions. However, this approach is challenged with the definition of a functional unit for comparison, the availability of comparable systems and the time demand to conduct the assessment (Kemna et al. 2005).

2.3.5 Findings

The pursuit of ecological sustainability through efficiency is considered as the most cost-effective and immediate strategy to reduce the demand of resources and associated emissions of activities. The adoption of the strategy to improve the energy demand of transformation processes raises the issue to quantify the actual and ideal energy performance in order to uncover the attainable potential for energy-related improvement. The quantification of improvement potential constitutes an integral element of (energy) performance management providing a consistent framework to derive information about the performance of activities and implement control mechanisms to pursue progress.

Revising the existing specifications and implications of energy performance, the following conclusions are drawn:

- Energy performance is an ambiguous term, which cannot be framed in one single, ideal indicator.
- The design and formulation of precise and verifiable indicators demands to consider the specific focus and purpose of the analysis as well as the context of the application.
- The specification of energy intensity as an energy efficiency indicator appears to be favourable for transformation processes underlining the usage of energy to pursue value creation.
- The setting of an ideal energy performance target remains the most critical aspect to retain the thrust towards improving the performance and bridge the performance gap. Yet, it is the fundamental element to trigger improvement.
- The measurement of performance and the target-oriented realization of improvement through management represent two elements conjointly forming a comprehensive energy performance management.

Chapter 3
Review on the State of the Research in Improving Energy Performance of Machine Tools

This chapter provides a review on the state of the research illustrating the status of current research approaches on improving the energy performance of machine tools. It begins with a short description of relevant approaches classified according to their specific focus within the process of energy performance management. An analysis of the approaches is then conducted evaluating the objectives and constraints in order to obtain a comparative overview about the state of research. In view of the evaluation of approaches, the further need for research is deduced to bridge the energy efficiency gap of machine tools.

3.1 Review on Existing Approaches

The energy demand of machine tools is affected by the material properties, selected process design and individual machine characteristics (Wolfram 1986). As a consequence, a variety of technical and organizational improvement approaches have been developed that specialize on specific origins of the energy demand. In accordance with the elements of performance management, the review on the state of the research comprises approaches to quantify, evaluate and improve the energy performance of machine tools.

The overview is structured in 6 sections providing an insight about the intention and ability to bridge the performance gap. First, descriptive approaches are delineated providing means to determine and quantify the electrical energy demand of machine tools before methods and tools for the evaluation of energy usage are reviewed. This encompasses specifically approaches that define minimum energy limits. In expansion to the first two sections, the third section revises approaches that are concerned with the integrated evaluation of energy usage extending the perspective towards technical and economic aspects. The improvement of energy performance through organizational or technical means is centred in scheduling and planning approaches from the production system level. Finally, available

A. Zein, *Transition Towards Energy Efficient Machine Tools*, Sustainable Production, Life Cycle Engineering and Management, DOI: 10.1007/978-3-642-32247-1_3,

approaches are analysed that provide a procedural concept including the measurement as well as the improvement of performance.

3.1.1 Descriptive Approaches

The estimation of the energy demand is a central element in the model of Schiefer to determine the environmental impacts of manufacturing processes (Schiefer 2000). The energy-related assessment encompasses the cutting energy to remove the material and the energy requirements to operate the machine tool (Schulz and Schiefer 1999). For the calculation of the cutting energy, the model requires information on the removed material volume and the force-induced cutting power, which is estimated using force prediction models (Schiefer 2000). The energy requirements of the machine tool are derived from the utilization time and the average power demands measured in idle as well as processing mode (Schulz and Schiefer 1998). The developed model obtains a product-oriented perspective estimating the effect of changes in product design on the energy demand of machine tools. The applicability of the model is bound to the availability of accurate cutting force models and the measurement of the specific power characteristics of the machine tool.

A comparable approach is pursued by Gutowski et al. describing the specific energy consumption (SEC) of machine tools as a functional relation based on the fixed power demand and the power requirement to process the material (Gutowski et al. 2006). While the fixed power P_{fixed} is anticipated as constant, the process power comprises the material removal rate υ as well as a material- and process-specific energy constant k (1) (Gutowski et al. 2007).

$$SEC = \frac{P_{fixed}}{\upsilon} + k \tag{1}$$

The description of the dependencies between the SEC and the material removal rate (MRR) enables to quantify the energy demand in relation to specific operating conditions (Gutowski et al. 2006). With regard to the absent specification of the material- and process-specific energy constant k, the elaboration of the model is initially unverified impeding the determination of the energy demand of machine tools.

Taking up the missing specification in the model of Gutowski et al., Li and Kara use design of experiments to characterize the relation between the energy demand and process parameters (Li and Kara 2011). The resulting empirical model quantifies the energy demand of specific machine tools as a function of the MRR with high statistical accuracy. It complies with the general structure of the model proposed by Gutowski et al. disproving however the presumption that the fixed power is a directly involved part of the model (Li and Kara 2011). The empirical modelling is applied for turning, milling and grinding processes on different machine tools providing different model coefficients for each observed case (Li et al. 2012).

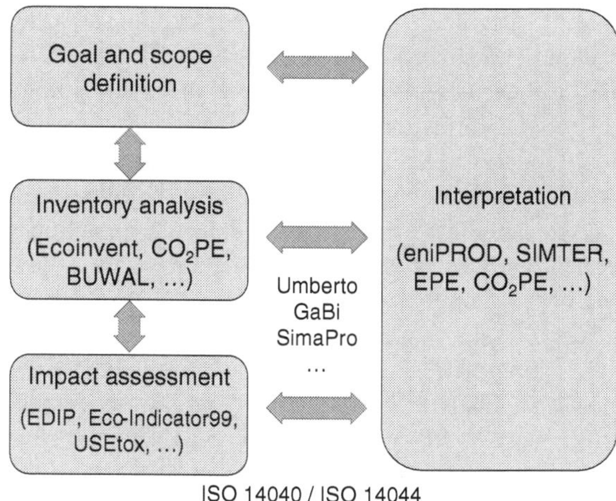

Fig. 3.1 LCA framework of the CO_2PE-initiative (Duflou et al. 2011)

These depend on the machine tool characteristics, process design and material properties (Kara and Li 2011). Therefore, the transferability of the empirical model for manufacturing processes is restricted necessitating measurements for each machine tool to specify the characteristic model. Diaz et al. take a corresponding approach relating the energy demand to the MRR (Diaz et al. 2011). An empirical model is derived in a case study of a micromachining centre, which confirms the general eligibility of the modelling approach to predict accurately the energy demand with respect to process parameters (Diaz et al. 2011).

Duflou et al. propose an inventorization method as part of the CO_2PE-initiative (Cooperative Effort on Process Emissions in Manufacturing) to gain information about the environmental impacts in the use phase of machine tools (Duflou et al. 2011). It inherits the procedure for conducting a life cycle assessment according to the ISO 14040 including the initial assignment of a scope of analysis, a functional unit and system boundaries (see Fig. 3.1) (International Standard Organization 2006; Kellens et al. 2011a).

The inventorization includes a screening approach and an in-depth analysis. While the screening approach strives to estimate the process and fixed energy demand using physical and functional approximations, the in-depth analysis considers time and power studies as a means to quantify the energy demand of machine tools in detail. In time studies, the operation of the machine tool is observed identifying functional modes and their time-shares during the period under study. The power demand in the different operational modes is recorded in power studies using measurement devices (Kellens et al. 2011a). Based on these datasets, the energy demand is calculated enabling the derivation of the energy-related environmental impacts. The interpretation phase encompasses finally the review of the results enabling the identification of improvement potentials (Duflou et al. 2011). The detailed analysis has been applied for a variety of processes starting with bending and milling (Devoldere et al. 2007), laser cutting (Devoldere et al. 2008), selective laser melting and sintering (Kellens et al. 2010) as well as electrical discharge machining (Kellens et al. 2011b). Although this methodology provides structural guidance on the determination of energy demands, the collection of datasets for machine tools appears to be limited by the anticipated time requirements of several hours to weeks for conducting the measurements (Kellens et al. 2011a).

Vijayaraghavan and Dornfeld delineate an automated monitoring system, which concurrently records the process data and power demand of manufacturing equipment (Vijayaraghavan and Dornfeld 2010). The approach eases consequently the effort to conduct time and power studies enabling an integrated data processing. By linking process and power data, events and operational modes that affect the power demand can be identified providing detailed insight into the processes of the observed system (Vijayaraghavan and Dornfeld 2010). The developed approach remains restricted to the introduction of the general system architecture. The functionality of the system is indicated in a case study for a machine tool providing initial information about the identification of events and reasoning.

Kuhrke et al. define a methodology to anticipate the energy demand already in the design stage of machine tools based on production profiles and power requirements of components (Kuhrke et al. 2010). A production profile encompasses estimations of the user about the utilization of the machine in productive and non-productive modes and anticipates functional requirements (e.g. activation of standby modes). By incorporating the power demands of components through specifications or measurements, the machine builder shall be enabled to predict the energy demand and improve the machine design for the given operational behaviour (Kuhrke et al. 2010). The proposed methodology underlines the demand to consider the operational behaviour of machine tools in the design phase and to minimize the effort to conduct power measurements. However, the approach remains limited to the representation of the methodology and does not provide an insight into the applicability and required data quality to predict accurately the energy consumption.

An approach to predict the energy demand of machine tools by aggregating the energy demand of components is pursued by Narita et al. as part of a concept for environmental assessment (Narita et al. 2006). It distinguishes a constant power demand for steady operating components and a dynamic power demand for the spindle and drives. The process-induced dynamics are derived deductively from cutting force models, which are used to predict the forces during the material removal, and provide an input to quantify the load torque affecting the power demand of the spindle and drives (Narita et al. 2003). As part of an overall machining simulation, the proposed model is applied in a case study enabling the quantification of energy demands for different machining strategies and process parameter variations (Narita et al. 2006). The elaboration of the proposed model is however bound to the estimation or measurement of the power demands of components and the specification of parameters in the partial force and torque models through empirical tests (Narita et al. 2003; Desmira et al. 2010). A corresponding approach is pursued by Eisele et al. modelling the power demand of machine components based on functional and physical relations. The proposed concept is exemplified for a supplementary coolant filter system for machine tools, yet presupposing the effort to specify the model parameters (Eisele et al. 2011).

Dietmair and Verl pursue a generic approach to estimate the energy demand modelling the machine tool operation as an arbitrary sequence of observed states with specific power demands (Dietmair and Verl 2009a). The discrete model encompasses states and transitions defining the linkages between states. A state depicts a specific power characteristic of a machine tool, which is assigned to a distinct functionality (e.g. machine ready for processing). The implementation of transitions enables to change the states using conditions, which are triggered through actions or timespans (Dietmair and Verl 2008). Accordingly, a model of a machine tool can be parameterized as sequential states through single power measurements (Dietmair and Verl 2009b). As the power demand of a defined state results from the accumulation of the individual power consumption for each component, the approach is extended towards a hybrid model structure taking first steps to integrate partial models. These enable to reconsider operational variations of components (Dietmair and Verl 2009a). A specification of the model is proposed using sub-models to predict the power demand of spindles and the power demand of coolant pumps (Verl et al. 2011; Dietmair and Verl 2010). The developed model framework allows forecasting the energy demand of existing machine tools, instantly providing the opportunity to integrate partial models of components for detailed analysis. The approach remains nevertheless constrained to conduct power measurements in order to derive the state/transition characteristics. It is moreover challenged to specify accurate parameters once sub-models are integrated to reflect process variations (Verl et al. 2011).

Schmitt et al. adopt the state-based modelling of Dietmair and Verl obtaining an entirely component-oriented perspective (Schmitt et al. 2011). The proposed methodology considers functional modes of machine tool components and the resulting power demands in order to depict the overall energy demand of the machine system. A new element of the approach is the definition of energy and time demands for transitions indicating the additional effort to switch between states (Schmitt et al. 2011). In contrast to the model of Dietmair and Verl, the proposed model remains limited to the representation of the conceptual structure and does not provide information about the anticipated implementation of the approach.

Larek et al. pursue a component-based modelling approach (Larek et al. 2011a) extending the scope towards the integration of data from the numerical control program of machine tools. Initially, the modes of components are extracted from present controls and included as data input in the model to reflect changes of modes (Larek et al. 2011a). The power demand is estimated as proposed in (Narita et al. 2006) allocating constant power demands for steady operating components and specifying the process-induced power requirements based on force-prediction models (Larek et al. 2011b). The applicability of the model to quantify the energy demand is demonstrated for a turning machining centre.

A similar concept is proposed by Avram and Xirouchakis obtaining data about the operational behaviour of components from the programmed codes of computer-aided manufacturing applications. The assessment of power demands for the spindle and drives is derived using prediction models for cutting forces and torques (Avram and Xirouchakis 2011). The estimation of the energy demand in four exemplary milling operations underlines the demand for accurate model parameters to predict precisely the power demand of the components.

3.1.2 Evaluative Approaches

Wolfram develops an approach to evaluate the operation of machine tools using the concept of embodied energy (Wolfram 1986). This includes the accumulation of the primary energy demands of the processed material and tool as well as the demands to operate the machine tool and supplementary systems in order to quantify the absolute energy demand of processing. The energy requirements of the machine tool are specified in detail using power measurements. By comparing the embodied energy of alternative machining strategies and machine tools, the developed approach demonstrates the ability to valuate holistically energy demands and identify improvement potential (Wolfram 1986). The applicability of the approach is nevertheless constrained in particular by the availability of contemporary data about the energy demands. Although Wolfram performs measurements to provide data for diverse machining processes, the transferability is limited due to a missing specification of the applied data acquisition procedures and aggregation. Moreover, the primary point of reference is set more than two decades ago providing a predominant proximity to manually operated machine tools.

Draganescu et al. focus on evaluating the efficiency η of energy usage in machine tools, which is defined as the indicator ratio of theoretical cutting power for material removal P_c to total power demand P_{total} (Draganescu et al. 2003). As the power demand is affected by the selection of process parameters, the functional relation is determined empirically for observed machine tools using statistical modelling of experimental data. Accordingly, the maximum obtainable efficiency of a machine tool is quantified as a function of the process parameters (Draganescu et al. 2003). Evaluating the efficiency solely based on the power demands is however misleading, as no information on the actual material processing and value creation is included. Draganescu et al. propose for that reason to consider the SEC, which integrates the cutting power P_c, machine tool efficiency η and material removal rate υ (2). In contrast to valuating the machine efficiency using the cutting power, the SEC is not referenced however towards an ideal efficiency level, which impedes a quantification of the obtainable improvement potential.

$$\begin{aligned} SEC &= \frac{P_{total}}{\upsilon} \\ &= \frac{P_c}{\eta}\frac{1}{\upsilon} \end{aligned} \tag{2}$$

As part of a framework development for the thermodynamic description of manufacturing processes, Branham et al. provide an exergy-based approach to assess the efficiency of material removal processes (Branham et al. 2008). The assessment considers the ratio of theoretical minimum work rate to actual utilizable exergy rate as efficiency indicator. Instantly indicating the challenge to define a minimum work criterion, possible approaches to define minimum work requirements are analysed. These encompass the work required to separate atomic layers, deform plastically a fracture of material or shear strains in metal cutting (Branham et al. 2008). Although the indicators are exemplarily calculated, the possible minimum work rates remain constrained to the presentation providing no information on the implications and conceptual meanings of the different efficiency indicators for manufacturing processes.

Renaldi et al. address the need for distinct efficiency indicators conducting a review on prevalent exergy efficiency indicators and their inherent informational value (Renaldi et al. 2011). By comparing the different indicator values for exemplary manufacturing processes, the dissimilarity in results is demonstrated emphasizing the necessity to select the indicator with regard to the intended scope of analysis. The use of isentropic efficiency indicators, which compare the actual exergy demand to an ideal, theoretical reference, is considered critically due to the problem of defining a minimum exergy value. For that reason, the application of exergy indicators based on input/output-analysis is proposed for the assessment of machine tools (Renaldi et al. 2011).

3.1.3 Integrated Evaluation Approaches

In addition to the specific assessment of energy usage, Schultz extends the scope of evaluation for manufacturing processes towards an economic and environmental perspective (Schultz 2002). An integrated process model is developed, which comprises the energy and material flows as an inventorization framework for the subsequent calculation of impacts. The environmental assessment is derived in accordance with the ISO 14040 integrating additionally a predefined weighting concept to aggregate the environmental indicator results into one comprehensive measure (International Standard Organization 2006; Schultz 2002). The quantification of the economic impact is derived based on an assignment of the direct and indirect costs to the prevalent energy and material flows. For the evaluation of manufacturing processes, the economic and ecologic measures are conjointly compared with reference entities or predefined target values enabling the quantification of improvement potential (Schultz 2002). The present approach establishes with the integrated process model a consistent basis for evaluation including structured acquisition and aggregation procedures for energy and material flows. As the subsequent valuation process focuses primarily on the comparison with alternative manufacturing processes or anticipated target values, it extends the valuation of energy toward planning. The approach does not provide information about the identification of an ideal reference object with minimum economic and environmental impact.

An alternative approach to evaluate manufacturing processes is pursued by Shin proposing a set of environmental and process-related parameters, which are

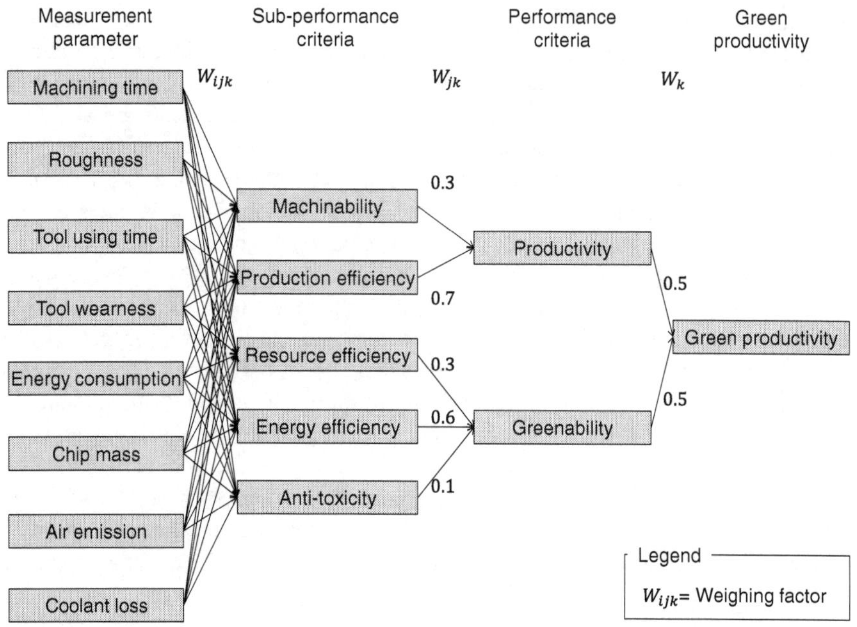

Fig. 3.2 Exemplary determination of a Green Productivity Index for machining (Shin 2009)

consolidated as visualized in Fig. 3.2 into an integrated Green Productivity Index (GPI) (Shin 2009).

The considered parameters encompass the energy demand as well as material and coolant losses as major factors contributing to the environmental performance (specified as greenability). Parameters affecting the productivity as second performance criterion are the process time, surface quality as well as tool operation time and wear. The conversion of the measured values to the Green Productivity Index is performed using matrices, which describe the relation between parameters and criteria based on expert knowledge (Shin 2009). In order to ease the effort of measuring the initial parameters, Shin extends the valuation concept with statistical models using experimental data to characterize the relation between process settings and performance measures. This enables to verify the effect of changing process designs on the GPI. Extending the valuation process of Schultz, the developed approach consequently provides insight on the attainable, process-related improvement potential for the observed machine tool. However, it has to be pointed out that the GPI is influenced especially by the weighted assessment of the measured parameters lacking a distinct, comprehensible prioritization.

Avram et al. develop an analogous methodology to evaluate machining processes aggregating technical, economic and ecological criteria through weighting factors into comprehensive measures (Avram et al. 2011). The approach defines 23 qualitative and quantitative criteria to assess the process design and machine tool. While the process perspective includes the cutting power as one of seven criteria, the machine level incorporates the idle power, total rated power of the machine and power demand to operate the spindle and drives (Avram et al. 2011). In contrast to the statistical model of Shin, the developed approach is constrained by the applied Analytic Hierarchy Process to consider solely known process alternatives and available machine tools. Consequently, an extensive number of experimental studies is required to generate case-specific data in order to determine an optimal reference system for the evaluation of improvement potential (Avram et al. 2011).

Anderberg et al. extend a machining cost model by integrating costs for tool wear and energy demand as well as anticipated carbon dioxide emissions in order to revise process parameters that minimize the overall costs (Anderberg et al. 2010). Due to the minor share of the energy-associated costs to the total machining costs, the optimal process rate is dominated by the direct machining costs and tool costs indicating the derogation between optimal economic and energy-oriented processing parameters.

Following the scope of the former approach, Rajemi develops a process optimization model adapting the general structure of economic optimization models for machining (Rajemi 2010). The approach considers the machine-related energy demand to initialize the processing, conduct the material removal, and change the tool. It integrates furthermore the embodied energy for the cutting tool. The approach is exemplified in a turning process identifying optimal processing parameters for different system boundaries. With regard to the considered system boundaries (e.g. excluding the energy demand for the tooling material), compliance between the purely economic and energy-oriented optimization is ascertained for an extended scope (Rajemi 2010). Although Rajemi describes the altering of the optimization results with varying system boundaries, a further elaboration of recommendations for the definition of boundaries is not provided.

3.1.4 Scheduling Approaches

In contrast to improving the process design, several approaches have been developed to minimize the energy demand of machining sequences by means of simulation and optimization. Mouzon determines multi-objective models to optimize the total energy consumption and completion time for sequences of machine tool operations (Mouzon 2008). The approach incorporates furthermore the Analytic Hierarchy Process to integrate preferences of decision makers and identify most appropriate out of available optimal solutions. From an energy-oriented perspective, the optimization of the sequencing enables to minimize the idle times of machines with high fixed power consumption and provides decision support to trigger a temporary shutdown and reactivation of machine tools between processes (Mouzon 2008). Rager provides an analogous concept to reduce the energy demand and associated costs through optimized scheduling (Rager 2008). The developed model focuses on the scheduling of identical, parallel machine tools to minimize the number of operating machines and extends the scope towards the levelling of energy demands to avoid costly energy peaks (Rager 2008).

Fang et al. address furthermore the minimization of energy demands through scheduling by introducing a model to optimize the cycle time, peak power consumption and carbon footprint (Fang et al. 2011). The model is exemplified in a milling case study indicating the trade-off between peak load of the machining operation and the cycle time. The approach expands the former optimization models by considering the adaption of process parameters (e.g. varying cutting velocity), instantly indicating the complexity of the optimization problem and compulsory computation time. Consequently, the application of the approach is impeded in industrial environments (Fang et al. 2011).

3.1.5 Planning Approaches from Production System Level

In addition to the minimization of energy demands through optimization, Weinert develops a methodology to improve the energy demand of manufacturing processes through simulation-based production planning and scheduling (Weinert 2010). The approach is based on process models, which depict the power demand of prevalent manufacturing processes as sequences of energy blocks (Weinert et al. 2011). These blocks constitute the specific power characteristics of operational states, which are parameterized incorporating process and machine properties. By integrating the energy blocks for production programs into simulation environments as *Plant Simulation*, the effect of different production strategies (e.g. minimize energy demand or balance machine utilization) on machine utilization, electricity consumption and electricity costs can be simulated to derive production programs (Weinert 2010).

Thiede provides an alternative approach developing a simulation environment to support the energy-oriented planning and control of manufacturing systems (Thiede 2012). With regard to the improvement of manufacturing processes, the developed simulation enables to predict the energy and resource demand of manufacturing processes as coupled elements in process chains with respect to process parameters and production programs (Herrmann and Thiede 2009). Using integrated process modules with measured power profiles or state-related power characterization, the effect of changes in process and machine design as well as scheduling production tasks on the energy and resource demand, associated costs and production output can be simulated enabling to improve the energy usage (Herrmann et al. 2011b).

3.1.6 Procedural Approaches

Binding formulates a systematic procedure to improve the energy and material demand of manufacturing processes and machined parts (Binding 1988). The comprehensive approach integrates four sequential steps starting with the quantification of prevalent energy and material flows, the assessment of improvement potential as well as the determination and evaluation of improvement options

(Binding 1988). Although each step of the procedure is elaborated in detail, the assessment of improvement potential remains constrained to the pairwise comparison of machined parts and machine tools. Binding provides neither an indication on how to determine efficient solutions as a reference objects nor an insight on the identification of suitable improvement strategies (Binding 1988).

A comparable approach is pursued by Engelmann developing a concept to deduce potential for the energy-related improvement of machine tools in the planning process (Engelmann 2009). The concept is subdivided into five steps starting with the estimation of process and energy requirements to perform the designated production process. These are determined based on the power characteristics of the machine tool and the anticipated operational behaviour. The identification of alternatives is described as a screening approach, which provides a set of generic improvement strategies to guide the identification of alternatives with improved energy usage. In the third step, the life cycle costs of all alternatives are assessed in order to determine the most economical solution (Engelmann 2009). While the first three steps strive to identify improvement potential in the planning phase, the remaining steps are concerned with the prevailing machine tool. Engelmann proposes the revision of planning data (e.g. the specification of functional requirements) after the machine installation as well as the integration of data into an energy process database as an expanding source of information for subsequent planning processes (Engelmann 2009). Striving to increase the awareness about the energy demand and associated costs of machine tools, the developed concept describes a general framework integrating the analysis and evaluation of energy demands of alternative machine tools (Engelmann et al. 2008). By proposing to review the general applicability of common improvement strategies, the approach does not specify and prioritize the identification of improvement potentials. Consequently, the concept restricts the guidance to realize an attainable improvement potential to the experience of the process engineer and the considered solutions.

3.2 Comparative Overview about the State of Research

In order to obtain a comparative overview about the state of research, the identified approaches to quantify, evaluate and improve the energy performance of machine tools are subsequently analysed and examined. This demands initially a clarification of classification criteria in Sect. 3.2.1 in advance to the assessment of approaches and deduction of further need for research.

3.2.1 Criteria for the Comparison of Approaches

Reflecting the requirements to bridge the energy efficiency gap for machine tools, the revised approaches are evaluated based on 19 attributes and characteristics, which are grouped into six main criteria categories.

- The *life cycle focus* reflects the emphasis of the approach towards a considered life cycle phase:
  - The design phase encompasses efforts within the sphere and responsibility of the machine tool manufacturer addressing energy-oriented enhancements of the machine tool structure and composition.
  - The use phase comprises all activities related to the machine tool operation that represent means to improve the energy demand.
- The *quantification* describes the method of determining information about the actual energy demand of machine tools:
  - Measurement facilitates capturing the specific energy requirements to operate present machine tools using power-metering devices.
  - Process models based for instance on cutting forces enable to predict the energy requirements for material removal with regard to varying process parameters.
  - *Component models* represent the basis to quantify the energy demand of machine tools aggregating the operational power demand of components.
- The *scope* specifies the intention and underlying overall goal of the approaches:
  - The *inventorization aims* at providing particular life cycle inventory datasets of unit processes to support the valuation of the environmental impacts through life cycle assessment.
  - Decision *support* relates to the identification of preferential choices among observed process or machine alternatives with regard to available information.
  - Process *improvement* reflects the attainment of energy-related saving potentials for machine tools through adjustments in the process of material removal.
  - Machine *improvement* includes all activities that are responsive to the energy-oriented enhancement of the machine tool system.
  - *Operational scheduling* is concerned with the balancing of machine tool operations in order to minimize the energy demand in non-productive modes.
- In order to guide the improvement of machine tools, particular approaches establish target values enabling the *evaluation* of energy performance:
  - *Target values* are used to indicate an aspired commitment as performance level based on predefined values.
  - The evaluation through *comparative assessment* relates to a determination of a relative improvement potential through monitoring or pairwise comparison.
  - *Theoretical limits* set performance thresholds based on thermodynamic and physical principles exceeding an attainable minimum for machine tools.
  - *Marginal values* represent an attainable energy threshold referencing a minimum energy demand based on practical solutions (e.g. best available technology).
- The *improvement* refers to the characteristics of means, which are applied to increase the energy performance:

- *Technical measures* aim at reducing the energy demand of machine tools through energy-improved technical alternatives (e.g. energy-saving valves).
- *Organizational measures* focus on saving energy through an improved control and operation of the machine tool system.
- *Optimization* refers to the identification and selection of optimal conditions to maximize the energy performance based on the formulation of an objective function.

The *implementation* refers to the applicability of the approach revising the capacity to ease the implementation and realization of improvement.

- *Procedural models* provide guidance on the initiation and operation of the approach.
- With regard to the organizational barriers impeding the improvement in energy performance, the approach incorporates methods and tools to ease the *effort* for the derivation and realization of improvement.

3.2.2 Comparative Overview

To review the results of the comparative assessment and indicate scopes with ensuing need for further research in a concise form, the revised approaches are displayed in matrices interconnecting the criteria (see Tables 3.1, 3.2). The compliance of each approach with the criteria is assessed according to three general characteristics such as "appropriate/considered" (●), "conditionally appropriate/considered" (◐), "not appropriate/considered" (○).

The results of the comparative assessment indicate that the dominating focus of research lies on the energy demand of machine tools in the economically and environmentally relevant use phase. With regard to the barrier of imperfect information, the necessity to quantify the energy demand is emphasized among all approaches. It depends essentially on experimental power measurements in order to determine the individual power characteristics of the observed machine tool. Striving to avoid the necessity of measurements, a minor share of the revised approaches partially integrates prediction models. These remain however constrained to a process or single component perspective necessitating to conduct power measurements nevertheless.

The main scope of approaches is concerned with the development of life cycle inventories in order to increase the availability and accuracy of datasets about the energy-related environmental impact assessment of machining processes (see Sect. 2.2.1). In addition to the inventorization, the enhancement of the energy demand through either process improvement or scheduling is of major concern. The improvement of the machine tool system is revised only in a small share of approaches maintaining a component-oriented perspective.

Despite the immanent need to quantify the energy demand, the energy-related evaluation of machine tools is only addressed in a marginal share of approaches.

Table 3.1 Comparative overview on the state of research

		Schiefer	Gutowski et al.	Li and Kara	Diaz et al.	Duflou et al.	Vijayaraghavan and Dornfeld	Kuhrke et al.	Narita et al.	Eisele et al.	Dietmair and Verl	Schmitt et al.	Larek et al.	Avram and Xirouchakis	Wolfram
Life Cycle	Design phase	○	○	○	○	○	○	●	○	●	●	●	●	●	○
Focus	Use phase	●	●	●	●	●	●	○	●	◐	◐	◐	◐	◐	●
Quantification	Measurement	●	●	●	●	●	●	●	●	●	●	●	●	●	●
	Process pred.	●	●	◐	○	◐	○	○	●	●	◐	●	●	●	●
	Comp. pred.	○	○	○	○	○	○	○	○	●	○	●	●	●	○
Scope	Inventorization	●	●	●	●	●	●	○	●	○	◐	○	○	○	○
	Decis. support	◐	○	○	○	○	○	○	○	○	○	○	○	○	○
	Process impr.	○	○	●	●	◐	○	○	○	○	○	◐	◐	◐	◐
	Machine impr.	○	○	○	○	◐	○	●	○	●	●	●	●	●	●
	Operat. sched.	○	○	○	○	○	○	○	○	○	○	○	○	○	○
Evaluation	Target values	○	○	○	○	○	○	○	○	○	○	○	○	○	○
	Comp. assess.	○	○	○	○	○	○	○	○	○	○	○	○	○	●
	Theor. limit	○	○	○	○	○	○	○	○	○	○	○	○	○	○
	Marginal value	○	○	○	○	○	○	○	○	○	○	○	○	○	○
Improvement	Technical	○	○	○	○	●	○	●	○	●	●	◐	●	○	●
	Organizational	◐	○	●	●	◐	○	●	○	○	◐	○	●	○	●
	Optimization	○	○	○	○	○	○	○	○	○	○	○	○	○	○
Implementation	Proced. model	◐	○	○	○	●	○	●	○	○	○	○	○	○	○
	Effort	○	◐	○	○	◐	●	◐	○	○	●	○	○	○	○

Table 3.2 Comparative overview on the state of research (continued)

		Draganescu et al.	Branham et al.	Renaldi et al.	Schultz	Shin	Avram et al.	Anderberg et al.	Rajemi	Mouzon	Rager	Fang et al.	Weinert	Thiede	Binding	Engelmann
Life Cycle	Design phase	○	●	●	○	○	○	○	○	○	○	○	○	○	○	●
Focus	Use phase	●	○	○	●	●	●	●	●	●	●	●	●	●	●	◐
Quantification	Measurement	●	●	●	●	●	●	●	●	●	●	●	●	●	●	●
	Process pred.	○	○	○	○	○	○	○	○	○	○	○	○	●	○	○
	Comp. pred.	○	○	○	○	○	○	○	○	○	○	○	○	○	○	○
Scope	Inventorization	○	●	●	●	◐	○	○	○	○	○	○	○	○	○	○
	Decis. support	○	○	○	●	●	●	○	○	○	○	○	●	●	●	●
	Process impr.	●	○	○	○	●	●	●	●	○	○	●	○	○	●	○
	Machine impr.	○	○	○	○	○	○	○	○	○	○	○	○	○	●	◐
	Operat. sched.	○	○	○	○	○	○	○	○	●	●	●	●	●	○	○
Evaluation	Target values	○	○	○	●	●	○	○	○	○	○	○	○	○	○	○
	Comp. assess.	○	○	○	●	○	●	○	○	○	○	○	○	●	●	●
	Theor. limit	○	●	●	○	○	○	○	○	○	○	○	○	○	○	○
	Marginal value	◐	○	○	○	○	○	◐	◐	○	○	○	○	○	○	○
Improvement	Technical	○	○	○	○	○	○	○	○	○	○	○	●	●	●	●
	Organizational	●	○	○	○	●	●	●	●	○	○	○	●	●	●	●
	Optimization	○	○	○	○	○	○	○	○	●	●	●	○	○	○	○
Implementation	Proced. model	○	○	○	●	○	○	○	●	○	○	○	●	●	●	●
	Effort	○	○	○	○	○	○	○	○	○	○	○	○	●	○	◐

The quantification of improvement potential in energy performance through reference values is sparsely included. It remains primarily limited to a pairwise comparison of alternatives or the definition of a process-oriented energy minimum. The formulation of comprehensive limits for machine tools remains disregarded.

In line with the dominating process perspective, the development of improvement measures is linked to organizational means determined to avoid especially the impact of the energy overhead not only in idle times. In contrast, technical measures as the substitution of components are described only in a minor share of the approaches.

To support the implementation of the developed approaches and methodologies in industrial environments, procedural models are provided in few approaches. In this context, the approach of Binding has to be pointed out providing a structured, comprehensive procedure to support the improvement of machine tools. The elaboration of the approach is however limited to a pairwise comparison of machine tools providing solely basic recommendations for improvement.

Despite the necessity to conduct for instance time-consuming power measurements or statistical analysis, the effort (indicated as time or costs) to implement and apply the approaches is only sporadically revised and concerned, thus endangering the practical appliance.

3.3 Ensuing Need for Research

Information and control are the two fundamental means to overcome the barriers, which obstruct the transition towards energy efficient machine tools. Reconciling the results of the review on the state of research with the requirements to bridge the energy efficiency gap, the following need for research can be derived:

The present state of research is characterized by the absence of a comprehensive, conceptual approach, which evaluates the energy performance of machine tools towards an unequivocal, ideal energy demand and provides systematic guidance for the design as well as operation of energy efficient systems. This involves initially the identification and methodological characterization of a reference threshold for the ideal energy demand as an evaluative means to quantify the maximum attainable improvement potential. The evaluation of energy efficiency of machine tools constitutes a prerequisite to exploit the technically attainable improvement potential. It necessitates essentially means to operationalize the potential into specific improvement policies. In mind of the organizational barriers obstructing the improvement of machine tools, all activities to evaluate and improve the energy performance need to be formalized and embedded in procedural models.

In line with the identified ensuing need for research, a concept for energy performance management is introduced and elaborated in the subsequent chapter enclosing the evaluation and improvement of energy efficiency to close the energy efficiency gaps of machine tools.

Chapter 4
Performance Management Concept to Evaluate and Improve the Energy Efficiency of Machine Tools

With regard to the ensuing need for research, this chapter introduces a comprehensive concept adapting performance management to evaluate and improve the energy efficiency of machine tools. First, the objective and functional requirements of the concept are presented, before an overview about the general concept layout and its structural elements is provided. Each element is then specified in detail illustrating the integrated methodologies and tools to determine an ideal energy performance, observe the actual performance and derive improvement. To guide and support the operation of the concept, the developed elements are coupled in workflow models formalising the processes to evaluate and improve the energy efficiency of machine tools.

4.1 Concept Objective and Preconditions

4.1.1 Objective

Revising the aspects that induce the energy efficiency gap for machine tools, an immanent need can be identified not only for technical and organizational improvement measures but also for comprehensive management concepts to overcome the barriers of imperfect information, profitability risks and organizational deficits. This research aims at the development of a performance management concept providing methods and tools to quantify the energy efficiency gap for machine tools and support the improvement towards an ideal energy level.

In this concept, the evaluation of energy performance is structured as a performance measurement process providing means to devise an ideal energy reference value and determine the actual performance through measurement. Based on the divergence in performance, the attainable improvement potential of machine tools is derived providing evaluative information in form of feedback and feedforward as part of the performance management process.

A. Zein, *Transition Towards Energy Efficient Machine Tools*, Sustainable Production, Life Cycle Engineering and Management, DOI: 10.1007/978-3-642-32247-1_4,

Striving to bridge the quantified energy efficiency gap for machine tools, the performance management concept inherits a systems perspective to support the analysis and planning of improvement measures for the machine tool. For this purpose, an analytical method is developed to disaggregate the energy consumption of machine tools into its system elements enabling a prioritization of energy consumers. In order to initiate improvement, an analysis of the relevant energy consumers against existing improvement measures is conducted as part of the concept. This facilitates the assignment of measures and development of improvement policies.

The transfer of the concept into prevalent management processes is exemplified using workflow models. These assist the integration of the developed concept into existing procedures and coordinate an effective application of the performance management process.

In respect to the identified demand for research, the performance management concept represents a comprehensive, systematic approach to identify and realize the attainable efficiency potential throughout the design and operation of machine tools. The application of the concept will provide particular encouragement to increase the efficiency of energy usage in machine tools and thus contribute to improve sustainability in manufacturing.

4.1.2 Functional Requirements and Preconditions

In order to ensure the attainment of the defined objective, functional requirements and preconditions are imposed on the development of the concept and its integrated process elements:

- **Energy performance management**: The process of energy performance management relies on the measurement of performance indicating the deviation between an ideal and actual performance. As the quantification of the improvement potential demands the definition of measurable performance indicators, the concept has to provide consistent means to derive reliable and valid metrics (see Sect. 2.3.2). This includes also the identification of an ideal energy performance as distinct and traceable reference value, which is attainable and yet flexible to adapt changes due to innovation and improvement.
- **Electrical energy demand**: The electrical energy demand of machine tools depends directly on the resistance to air flows, inertia of masses and especially the technology specific friction that comprises the machining operation (Bartz 1988). Establishing an ideal energy requirement has therefore to take into account the technological characteristics and maintain essentially a specific scope for each machining technology to be observed in order to ensure the comparability for the performance evaluation.
- **Metering strategy**: As the determination of the actual energy performance requires to conduct power measurements, a metering strategy has to be derived specifying the parameters and conditions for the energy assessment of machine

tools. Using a unified procedure ensures a consistent and effective acquisition of data about the energy demand enhancing as a result the reproducibility of measured results.

- **Improvement potential**: The electrical energy demand of machine tools is affected by a variety of measures starting from the process design and material specifications to the machine configuration (Wolfram 1986). Revising the strong focus of research on the process improvement in the use phase (see Sect. 3.2.2), the developed concept has to be extended to conjointly evaluate the process and machine potential for the design of new and the operation of existing machine tools.
- **Measures**: As indicated in Sect. 2.2.3, a broad scope of technical and organizational measures is increasingly available addressing specific aspects of energy usage. In order to capture the full potential for improvement, the concept under development has to take a systems approach endeavouring the enhancement of the entire machine tool system and not remain constrained to isolated machine elements.
- **Management process**: Improving the energy efficiency of machine tools through performance management requires a structured, systematic management process. This needs to be addressed through the design and implementation of workflow models, which support the application of the methods and tools in effective business processes to evaluate and improve the energy performance.

4.2 Concept Layout and Structure

In line with the formulated objective to develop a performance management concept for the evaluation and improvement of the energy performance of machine tools, a conceptual layout with its integrated modules is developed depicting the envisaged methodologies, tools and processes. The general outline is illustrated in Fig. 4.1.

The concept is composed of four concept modules to identify the ideal energy performance (1), to determine the actual performance (2), to analyse improvement potential (3) and to plan improvement opportunities (4) for the integrated energy consumers in the machine tool system. For the operation of the modules, workflow models are defined guiding the process to evaluate and improve the energy efficiency (5).

- The first module aims at the **determination of an energy performance limit (1)**. In accordance with the process of performance measurement, it includes initially the specification of an energy limit in form of quantitative metrics. For the identification of the ideal energy performance, a methodological approach based on production frontiers is introduced to derive an objective, realistic and challenging reference for the performance comparison. The specification of the

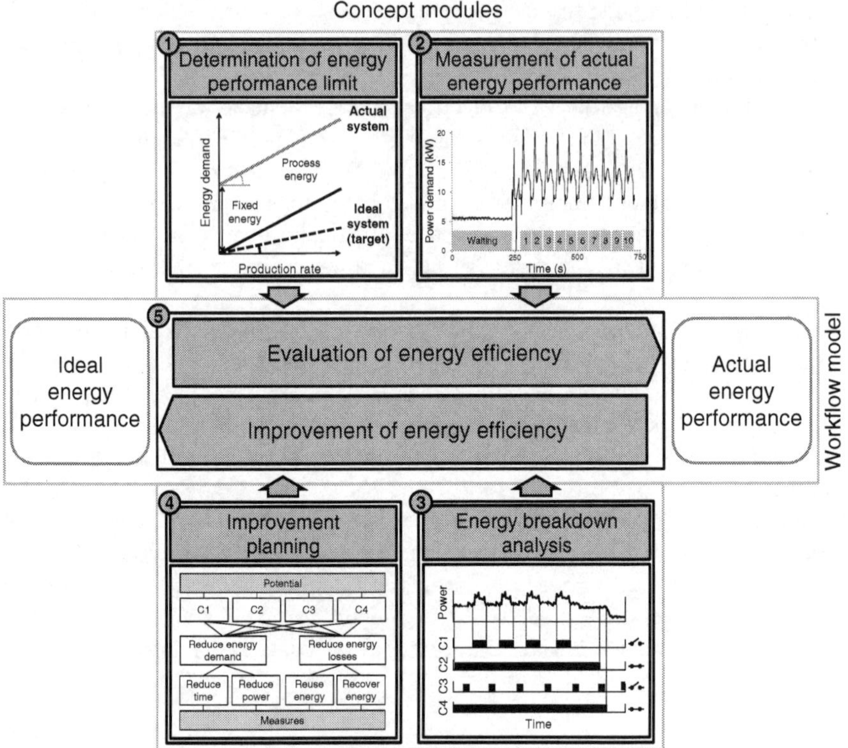

Fig. 4.1 Structure of the performance management concept

production frontier for machine tools is summarized providing guidance for the formulation of energy limits for different machining technologies.

- The second module defines procedures to *measure the actual energy performance (2)* of a machine tool under study. Based on the definition of a unified metering strategy, the power demand of the machine tool is measured during defined idle mode and material removal sequences in order to derive a consistent power characteristic of the system. By converting the measured power demands into indicator values, the actual performance is determined for the specific machine tool as an input for the quantification of the energy efficiency gap.
- With regard to the diversity and heterogeneity of machine tool configurations and operations, the module on *energy breakdown analysis (3)* provides a universal methodology to ease the disaggregation of the total energy demand and the detection of the origins of energy usage in the machine tool system. The methodology provides therefore means to record concurrently the operational machine data and power demand. The resulting dataset supports consequently the need to understand the power demand with regard to the operational characteristics of the integrated machine tool components. By allocating and

prioritizing the energy demand of components, the energy breakdown analysis constitutes the basis for the identification of improvement opportunities based on the temporal operation and power demand of the machine tool elements.

- The *improvement planning (4)* represents a concept module to allocate improvement measures to potentials providing the basis for the formulation of improvement policies. It considers technical as well as organizational improvement measures for the design and operation of machine tools striving to increase the actual performance towards the ideal reference value. Linking improvement measures according to the energy performance limits, the developed module provides a consistent routine enabling a structured review and assignment of measures. Based on the selection of measures, the attainable saving potential for each measure can be calculated specifically for the analysed machine tools in order to define a set of cost-effective improvement actions.
- The *workflow model (5)* for the evaluation and improvement of energy efficiency structures the application of the proposed concept modules *(1–4)* for the definition of energy performance limits and measurement using workflow models. These ensure an effective integration of the processes by defining activities and actors involved. As a result of the evaluation process, the attainable improvement potential is quantified completing the process of performance measurement. Based on the identified divergence between the actual and ideal performance, the improvement process is initiated triggering the enhancement of the actual performance value. This includes the operation of the concept modules to revise specific saving opportunities and to associate measures for the design and operation of components. The activities within workflow models are equipped with feedback mechanisms in order to implement an iterative operation of the evaluation and improvement process. The evaluation and improvement processes are employed iteratively in order to establish a continuous improvement process.

Each module of the proposed concept for energy performance management is elaborated in the following sections starting with the concept modules for the evaluation and improvement of energy performance in advance to the formulation of the workflow model.

4.3 Determination of an Energy Performance Limit

4.3.1 Energy Performance Limits

The formulation of a target value is a central element in performance measurement expressing the commitment to attain a designated level of performance (Stoop 1996). As indicated in the state of research, there is no generally accepted methodological approach available to define and quantify an energy performance limit for machine tools.

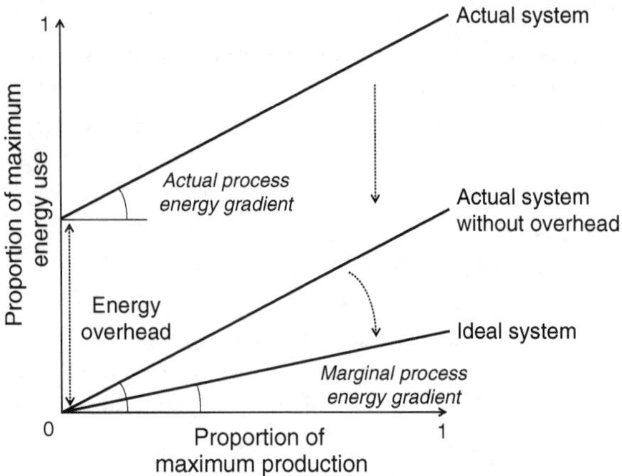

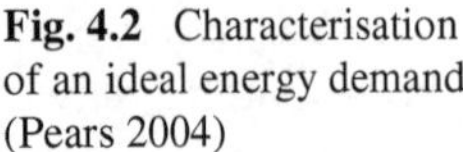
Fig. 4.2 Characterisation of an ideal energy demand (Pears 2004)

Pears provides a generic perception of an ideal energy performance for transformation processes reviewing the energy demand as a function of the production output from a systems perspective (Pears 2004). As illustrated in Fig. 4.2, the energy demand of an ideal system is characterized by two properties (Pears 2004):

- The transformation process does not feature a fixed energy overhead, thus requiring no energy in non-operating times.
- Secondly, the gradient of the marginal process energy reflects the minimum amount of energy to conduct the transformation.

While the first property regarding the energy overhead represents a quantifiable metric and can serve as a target for the performance measurement process, the marginal process energy remains unspecified and requires consequently further clarification. Both properties together form a basis to deduce energy performance limits for machine tools.

The determination of the marginal energy gradient as a target value is challenged to find a useful interpretation of corresponding energy limits and an associated method to quantify the energy minimum. Descriptions of energy thresholds vary with regard to the assumptions and underlying estimates from theoretical to market induced minimum energy demands (see Fig. 4.3) (Jaffe 1994; Koopmans and te Velde 2001). The *theoretical energy limit* relates on thermodynamic and physical principles indicating the energy demand to perform the transformation (e.g. material removal mechanisms) (Reap and Bras 2008; Seryak and Kissock 2005; Takada et al. 2000). Apart from the theoretical interpretation, a practical energy limit considers additionally energy to cover tribological losses in form of friction and wear (Bartz 1988). The term *best available technology* indicates in this sense a practical energy minimum, which can be achieved using the most effective and advanced technology available to conduct the transformation (Organisation for Economic Co-operation and Development 1999; Mulvaney 2011). It is therefore considered as a realistic and attainable threshold of a

Fig. 4.3 Forms of energy process limits

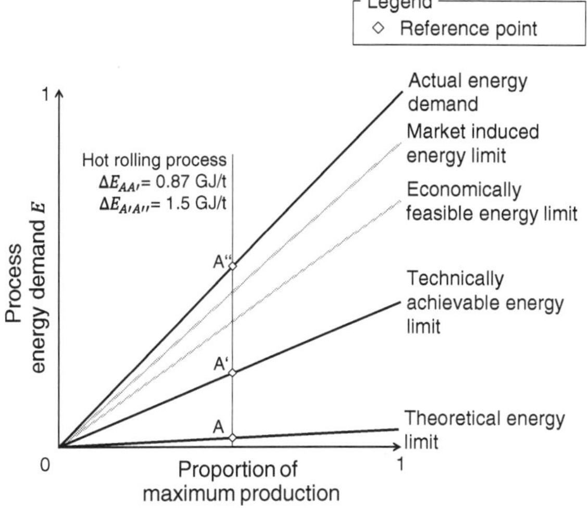

marginal energy demand (Takada et al. 2000). By distinguishing the *technically achievable* and *economically feasible energy limit*, the consideration of profitability is included affecting the magnitude of the energy limit based on the evaluation of energy savings and associated costs (Seryak and Kissock 2005). The *market induced energy limit* describes the minimum energy demand, which results from imperfect information about the availability of economic improvement measures (Jaffe 1994).

The results of a theoretical analysis about the energy demands of steel making processes provide an initial insight and better understanding about the magnitudes between the energy limits. In the example of a hot rolling process, the theoretical minimum energy demand A' and technically achievable minimum demand A' are estimated with 0.03 and 0.9 GJ/t. The energy demand of an actual hot rolling process A'' is furthermore quantified with 2.4 GJ/t (Fruehan et al. 2000). While the potential of 0.87 GJ/t corresponds to a theoretical option, the potential of 1.5 GJ/t in the second case represents an actually achievable saving capacity (see Fig. 4.3).

To support the determination of a suitable performance target, general requirements for performance measurement have been formulated in order to evade risking failure of the improvement process and ensure a vital as well as successful realization (Stoop 1996; Johnston et al. 2001). The requirements on performance targets comprise the definition of an objective, realistic and challenging reference value (Stoop 1996). From the identified energy thresholds, the technically achievable energy limit is chosen for the development of the concept module. As an energy performance target, it provides the best correspondence with the defined requirements. In addition, it is universally applicable and avoids coping with variances in the economic consideration of energy savings, which affect the evaluation of economic feasibility.

Following the specification of an energy performance limit, the developed concept module is relying on a methodology to determine the technically achievable energy minimum for machining processes. For that reason, the energy demand of machine tools is reviewed in the following section from a production theoretical perspective revising the feasibility to interpret the production frontier as best available technology (Fandel 2010). As a quantitative model, the production frontier ensures furthermore consistency with the requirements to define target values for performance measurement (as indicated in Sect. 2.3.2) (Seidl 2002; Stoop 1996; Sturm 2000).

4.3.2 Performance Analysis of Transformation Processes using Production Frontiers

The review and analysis of transformation processes by observing the input and output flows as well as their relationship is of fundamental concern to production theory (Koopmans 1951; Dyckhoff 2006). It provides deterministic solutions for deductive reasoning and serves the function to describe, explain, predict and direct the transformation for value creation by means of decision and control as well as communication and learning (Koskela 2000; Krause and Mertins 2006; Dyckhoff 2006).

Within the scope of production theory, the performance of a transformation process is generally assessed based on the quantitative ratio of output to input, which is referred to as *productivity metric* (5) (Coelli et al. 2005). It represents a descriptive performance indicator of a transformation reflecting the demand of resources to create a distinct output (Dyckhoff 2006; National Research Council 1979). When considered individually, productivity does not provide any use as a single performance measure. Only through comparison with corresponding productivity measures, qualitative information on the performance of the transformation can be obtained (Cantner et al. 2007).

$$Productivity = \frac{Output}{Input} \tag{4.1}$$

Revising the definition of energy efficiency indicators (see Table 2.3), the production theoretical formulation of productivity for energy usage (energy productivity) does comply with the quantitative ratio of energy intensity as a technically oriented metric (Diekmann et al. 1999).

All productivity ratios of activities, which can be realized through input–output combinations, form a *feasible production set of a technology* (Coelli et al. 2005). The upper bound of the respective production set constitutes the dominating transformations of a technology describing the relationship to attain the maximum output from a given input (Koopmans 1951). In accordance to the economic principle, the dominating activities define efficient transformations enclosing the considered technology as *production frontier* (see Fig. 4.4) (Dyckhoff and Spengler 2010; Coelli et al. 2005).

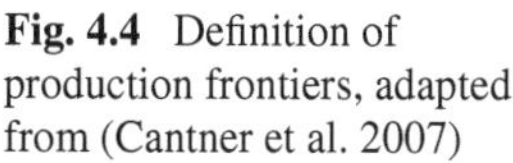

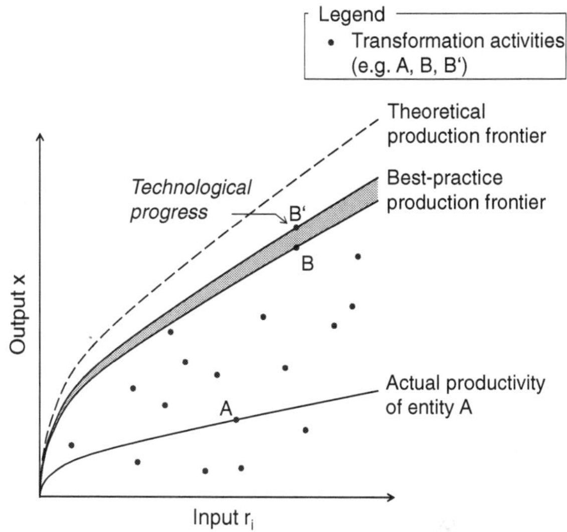

Fig. 4.4 Definition of production frontiers, adapted from (Cantner et al. 2007)

In practical applications, the production frontier is commonly uncharted due to imperfect information or operational constraints (Kleine and Dinkelbach 2002). To cope with this limitation, approaches have been developed to approximate the production frontier through empirical analysis of observed entities (Zhu 2009; Allen and Dyckhoff 2002). The resulting functional relation describes in this way the current technically achievable productivity of a technology. It classifies efficient and inefficient activities based on the respective productivity values and serves as a benchmark of the technologically available possibilities (Kumbhakar and Lovell 2003; Allen and Dyckhoff 2002). The production frontier facilitates in this way to guide development towards best practice solutions (Bogetoft and Otto 2011). As illustrated in Fig. 4.4, it enables furthermore to monitor technological progress, pushing the boundary of the frontier towards the theoretical limit (Fandel 2010).

The approaches to estimate the production frontier can generally be distinguished according to the underlying assumptions on the stochastic behaviour and the functional form of the transformation (see Table 4.1) (Bogetoft and Otto 2011; Dyckhoff 2006). The corrected ordinary least squares (COLS) model is an example of a deterministic and parametric approach. It determines an ordinary

Table 4.1 Taxonomy of methods to approximate the production frontier (Bogetoft and Otto 2011)

		Measurement relative to frontier	
		Deterministic	Stochastic
Estimation method	Parametric	Corrected ordinary least squares (COLS)	Stochastic frontier analysis
	Non-parametric	Data envelopment analysis (DEA)	Stochastic data envelopment analysis (SDEA)

regression model as a parametric relation, which is shifted parallel towards the efficient activity dominating all other transformations (Bogetoft and Otto 2011). An alternative parametric means is the stochastic frontier analysis (SFA), which extends the identification of a functional relation with error terms to measure the inefficiency and to account for random noise (Coelli et al. 2005; Aigner et al. 1977). It enables consequently to extract disturbances, which affect the outcome and are not in control of the transformation (Greene 2008). In contrast to the parametric approaches, the data envelopment analysis (DEA) determines the production frontier without formulating extensive assumptions on the properties of the technology. As a non-parametric and deterministic model, it compares basically the productivity of observed entities in order to enclose the efficient alternatives with a gradually enveloping linear contour (Coelli et al. 2005; Greene 2008). An additional modelling approach to estimate the production frontier is the stochastic data envelopment analysis (SDEA). The non-parametric and stochastic model extends the scope of DEA by accommodating stochastic influences and random noise affecting the data measurement (Bogetoft and Otto 2011; Ray 2004).

Among the revised approaches, the DEA and SFA represent the commonly used modelling approaches to depict empirically the production frontier (Coelli et al. 2005). With regard to the underlying assumptions and the consideration of outliers as stochastic errors, both approaches provide deviating results in terms of the production frontier (Burger 2008). The merits of the SFA are attributed particularly to the consideration of stochastic effects and the ability to perform statistical testing in order to verify the significance of the functional relation. In contrast, the DEA surpasses the SFA imposing no obligation to estimate a functional form of the production frontier and comparing the performance towards identified transformations rather than statistically aggregated measures (Coelli et al. 2005; Bogetoft and Otto 2011). The selection of methods depends on the specific requirements and in addition on the availability of data to describe the transformation process (Coelli et al. 2005; Burger 2008). A comprehensive overview about the properties and applicability of approaches is provided in Andor (2009), Bogetoft and Otto (2011), Burger (2008) and Coelli et al. (2005).

Based on the approximation of the best-practice production frontier as a reference of efficient transformations, the comparison of the productivity between an actual entity and the frontier enables to compute a specific *degree of efficiency*. It is a relative, quantifiable metric indicating the divergence between the two productivity levels (Dyckhoff 1994). The degree of efficiency is also referred to as *technical efficiency*. It quantifies the inefficiency in form of an improvement potential to increase the output or minimize the input (Färe et al. 1985; Coelli et al. 2005). In order to maintain consistency throughout the quantification of inefficiency, the distance function to the efficient frontier and the measurement procedures to conduct the observation need to be defined in advance. In the case of one input and one output, three options can be defined as technical efficiency measuring the distance between the inefficient activity and the efficient frontier of the technology (see Fig. 4.5) (Allen and Dyckhoff 2002).

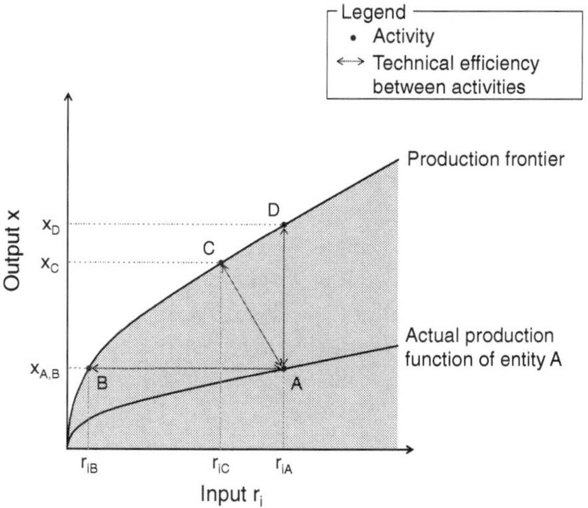

Fig. 4.5 Approaches to determine the technical efficiency based on the production frontier, adapted from (Coelli et al. 2005; Cantner et al. 2007)

The input-oriented measure of technical efficiency (indicated as *AB*) relates to the minimization of input resources to create a constant amount of output. The output-oriented scope (*AD*) inverts the consideration by increasing the output for a given set of inputs (Coelli et al. 2005). A third option (*AC*) measures the inefficiency as the minimal distance between the production frontier and actual input–output combination of the entity indicating the simultaneous adaption of input and output (Allen and Dyckhoff 2002).

The empirical approximation of the production frontier provides a methodological means to observe a practical benchmark of a technology enclosing the technically achievable best-practice solutions. The interpretation as a performance target suits the function to guide improvement through quantifying the inefficiency and to monitor technical progress (Bogetoft and Otto 2011; Zhu 2009).

4.3.3 *Specification of Energy Production Frontiers for Machine Tools*

An initial step in the empirical approximation of a production frontier for energy usage in machine tools is to decide on an estimation methodology. To specify an energy production frontier through parametric approaches, a functional form is presupposed depicting the interdependence between the energy demand and material removal (as reference to a machined part) (Fandel 2010). A functional relation for the transformation of resources in machine tools can be deduced either through inductive or deductive reasoning (Hill 2011; Haberfellner 1999).

The *engineering production functions* represent an example of an inductive approach. The functional relation between input and output is inferred from a set of disaggregated process models (Sonntag and Kistner 2004). These describe the elementary operations to convert energy in transformation systems following physical and chemical principles (Fandel 2010). A particular advantage of the engineering production functions is the attainable accuracy of the functional relation, which is interdependent with the reliability of the included models (Fandel 2010). As indicated in the state of research, the availability and applicability of detailed elementary models describing the individual energy conversion in machine tools is limited (see Sect. 3.1.1). The lack of models thus constitutes a barrier, which constrains the continuation of this approach.

A representative of the deductive modelling approaches is the *Gutenberg production frontier*. Deductive approaches express a functional relation in form of a predefined theory or generic model, which is adapted to reflect the individual characteristics of the transformation (Dyckhoff 2006). It facilitates the determination of the specific functional form for a transformation system under study and avoids conducting detailed technical analysis (Fandel 2010). The Gutenberg production function is concerned in particular with the influence of industrial transformation systems on the conversion of inputs and outputs. It is based upon consumption functions, which indicate the technical capability of the operating equipment (Sonntag and Kistner 2004).

In order to assess the suitability of the Gutenberg production frontier as a parametric approach to estimate the energy production frontier, the energy usage in machine tools is subsequently modelled according to the underlying functional relation.

4.3.3.1 Modelling the Energy Usage in Machine Tools

The Gutenberg production frontier provides a methodological approach to describe the conversion of resources in industrial transformation systems distinguishing consumption and usage factors (see Fig. 4.6) (Fandel 2010; Dyckhoff and Spengler 2010).

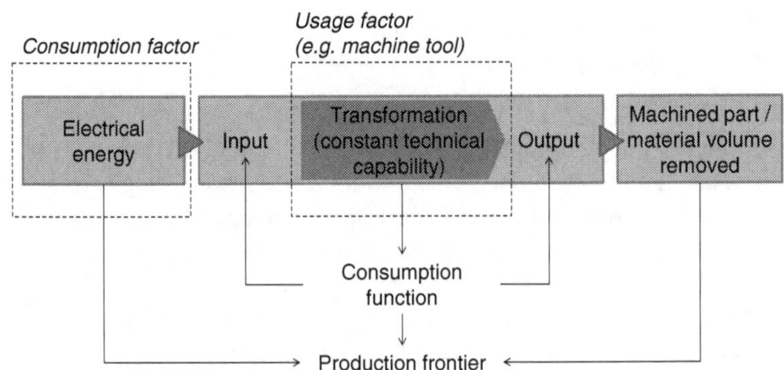

Fig. 4.6 Elements of the Gutenberg production frontier

Extending the scope of the Leontief production function, the Gutenberg production frontier disengages the prevalent constant relation between an input r_i and output x (as assumed in input–output analysis) (Dyckhoff 2006). It formulates instead a variable consumption function $a_i(\lambda)$ for each consumption factor i under study (4.2) (Fandel 2010). The function internalizes the individual characteristic of the conversion of input into output. It indicates in this way the technical capability of the usage factor to perform the transformation (Albach 1980; Dyckhoff 2006). As a control variable, the performance intensity λ represents an adjustable operating point of the usage factor, which determines the associated resource demand on the basis of the consumption function (Dyckhoff 2006).

$$r_{in} = a_{in}(\lambda_n)\,x \qquad (4.2)$$

$$i = 1, \ldots, I, \qquad n = 1, \ldots, N$$

In general, a set of consumption functions can be formulated reflecting the technical capability z for each considered consumption factor i and the performance intensity λ of the usage factor n (4.3) (Fandel 2010). The consumption function is constrained by the observed technical characteristics of the usage factor. Any change affecting the usage factor and its technical capability exerts therefore influence on the relation between input and output demanding to renew the consumption function (Sonntag and Kistner 2004; Adam 2001). Due to technical constraints and quality claims on the transformation output, the scope of the consumption function can be bound by limits on the performance intensity enclosing a process window of feasible transformations (Sonntag and Kistner 2004).

$$a_{in} = a_{in}(z_{1n}, \ldots z_{Vn}, \lambda_n) \qquad (4.3)$$

$$i = 1, \ldots, I, \qquad n = 1, \ldots, N, \qquad \underline{\lambda_n} < \lambda_n < \overline{\lambda_n}$$

In Fig. 4.7, the consumption functions of three input resources are exemplified for a drilling process of a single part under the assumption of a constant technical capability. The functional relations are displayed relative to the revolutions per minute (rpm) of the drilling head as intensity metric. In this way, the consumption function enables to model the constant demand on cutting fluids (if decoupled from the operational mode and output), the diminishing labour demand as well as the fixed and process-related energy characteristics with increasing processing speed (Dyckhoff 2006). It is therefore possible to anticipate the effect of varying performance intensities and identify an ideal operational mode with minimal resource demand (Sonntag and Kistner 2004).

The primary approach to derive consumption functions is the empirical observation. This encompasses laboratory experiments by the equipment manufacturer or measurements during the use phase of the equipment (Fandel 2010). The general suitability to incorporate empirically observed functional relations for industrial processes into the concept of the Gutenberg production frontier has been verified in several empirical case studies ranging from chemical to metalworking processes. To authenticate the compliance of observed relations as consumption functions, two criteria are specified as follows (Fandel 2010; Albach 1980):

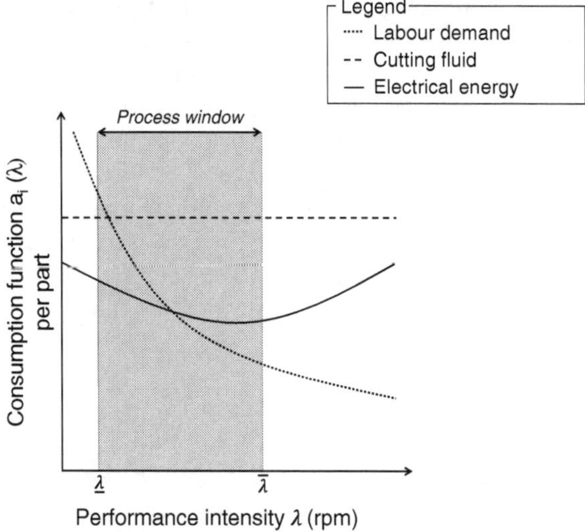

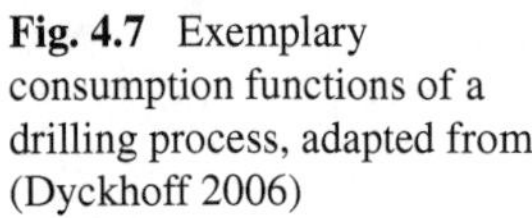
Fig. 4.7 Exemplary consumption functions of a drilling process, adapted from (Dyckhoff 2006)

- The output x of a transformation is expressed as a function of the performance intensity λ and the time demand for processing t.
- For a defined performance intensity λ and constant technical capability z of the usage factor, the consumption function $a_i\ (\lambda)$ has to provide an unambiguous relation with limitational factors between the input r_i and output x.

To review the implications of the imposed criteria on the formulation of an energy consumption function a_E, the functional relation describing the energy demand of machine tools is empirically analysed for an exemplary grinding process.

Using design of experiments, the energy demand to operate an internal cylindrical grinding machine tool has been determined in power measurements with three-phase power meters for a defined set of process parameter constellations (Herrmann et al. 2009). The selection of process parameters includes the material removal rate (MRR) υ, the cutting velocity σ and the removed material volume ρ as controllable variables with individually specified process windows (see Fig. 4.8). For machining processes, the removed material volume can be used as a suitable proxy indicating the amount of machined parts of the transformation process. The processing material, tool and cutting fluid remained unchanged throughout the experiments. The energy demand to conduct the material removal is captured as single measure and accumulating the power consumption during the cutting period. The effect of altering process parameter settings on the energy demand is determined through statistical analysis with the software application Design Expert®. It derives a functional relation between the controllable variables characterizing the energy consumption for material removal (Antony 2010; Herrmann et al. 2009).

Based on the results of the grinding experiments, the statistical analysis revealed a functional relation between the energy input and material output in

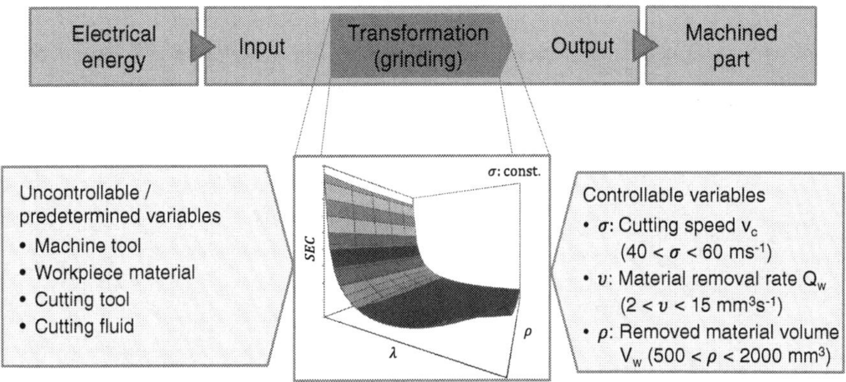

Fig. 4.8 Energy-related design of experiment of a grinding process

form of a quadratic regression model (4.4). In the interest of clarity, the statistical model and the units of the individual constant factors and parameters are provided in the Appendix B.

The outcome of the statistical analysis underlines a strong dependence between the MRR and the energy demand (Herrmann et al. 2009). In case of a constant material removal volume and cutting speed, the energy demand can be expressed exclusively as a function of the MRR (4.5). The resulting function is visualized in Fig. 4.9.

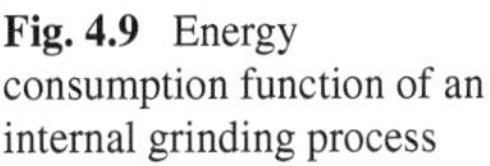

Fig. 4.9 Energy consumption function of an internal grinding process

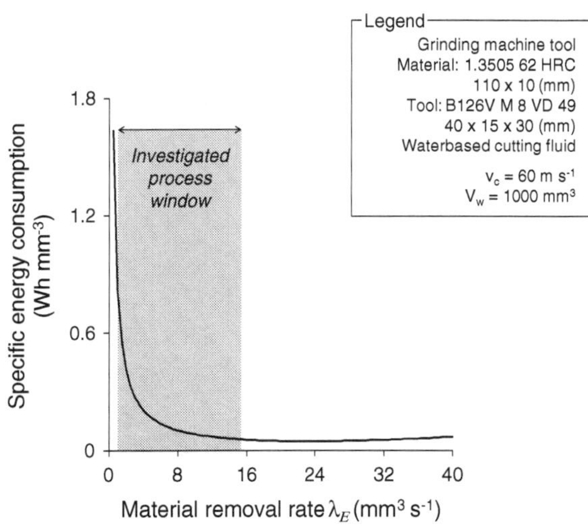

$$a_E(\sigma, \rho, \upsilon) = 25.39 - 0.38\sigma + 0.009\rho + 0.81\rho\upsilon^{-1} - 2.77\upsilon + 0.03\sigma\upsilon - 0.0005\rho\upsilon + 0.06\upsilon^2 \quad (4.4)$$

$$R^2 \text{ adjusted} = 92.53\,\%,$$
$$\underline{\sigma} < \sigma < \overline{\sigma}, \underline{\rho} < \rho < \overline{\rho}, \underline{\upsilon} < \upsilon < \overline{\upsilon}$$

$$a_E = (\upsilon) = 10.87 + 813.21\upsilon^{-1} - 1.31\upsilon + 0.06\upsilon^2 \quad (4.5)$$

$$\sigma = 60\text{ ms}^{-1},\ \rho = 1{,}000\text{ mm}^3,\quad \underline{\upsilon} < \upsilon < \overline{\upsilon}$$

Assigning the MRR υ as performance intensity metric λ, the derived functional relation in (4.4) fulfils both compliance criteria. It expresses the output of the transformation based on the operating time and the material removal rate. The derived relation provides additionally a distinct functional relation between the energy requirements and removed material volume for a given intensity and technical capability. The quadratic regression model can therefore be considered as consumption function of the Gutenberg production frontier with the MRR as energy performance intensity metric λ_E. Accordingly, any functional relation, which models the energy demands of machine tools relative to the MRR, can be interpreted as energy consumption function a_E within the Gutenberg production frontier.

On the basis of the energy consumption function, the energy requirements and removed material volume are quantified for varying performance intensities. Each level of performance intensity establishes an unambiguous relation between the input and output. These relations define individual, linear production functions for the observed machine tool. An extract of the resulting array of production

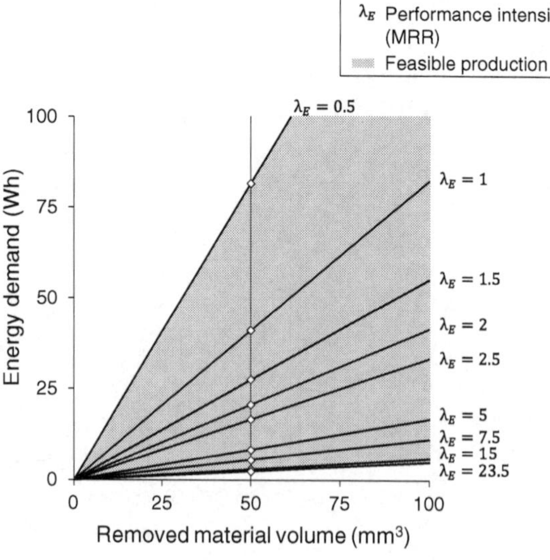

Fig. 4.10 Selection of production functions for the grinding machine with varying performance intensity

functions for the grinding machine is visualized in Fig. 4.10. The feasible input–output combinations form an associated production set indicating the different energy requirements of attainable activities to process a certain output quantity (Albach 1962). Comparing the corresponding productivity measures of each production function, the dominating transformation of the production set represents the production frontier for the machine tool under study (Albach 1962).

Characterizing the conversion of energy for material removal as energy consumption function, the determination of the production frontier takes specifically into account the technical capability and processing performance of the machine tool under study. The influence of both factors on the energy productivity of machine tools is subsequently considered in detail. Extending the scope from a single machine tool with constant technical capability towards a machining technology, an approach to approximate the enclosing energy production frontier with variable technical capability is introduced. It pursues the identification of benchmarks among machine tools with minimal energy requirements ascertaining instantly a target value for the energy performance measurement.

4.3.3.2 Approximation of the Energy Production Frontier for Machine Tools

The estimation of an energy production frontier relies on the identification of efficient transformations. A transformation is considered as efficient, if the productivity is not dominated by another transformation generating either the same output with less input or more output with equal input (economic principle) (Koopmans 1951; Cantner et al. 2007).

Modelling the relationship between the energy demand and material removal in form of the Gutenberg production frontier, the determination of productivity (4.4) can be transposed from the assessment of input–output ratios to an evaluation of the energy consumption function (4.6) (Fandel 2010). In this context, the productivity value takes not only the energy demand for material removal into account but also the particular performance of the machine tool. It provides therefore the opportunity to compare machine tools with different technical capacities at a given performance intensity (in form of the MRR).

$$P_E = \frac{Output\ x}{Input\ r_E} = \frac{\lambda_E t}{a_E(\lambda_E)\lambda_E t} = \frac{1}{a_E(\lambda_E)} = P_E(\lambda_E) \tag{4.6}$$

$$\underline{\lambda_E} < \lambda_E < \overline{\lambda_E}$$

The influencing factors on the energy productivity P_E are the performance intensity and technical capability of the machine tool (4.5). While the performance intensity maintains a process-oriented perspective, the technical capability is concerned with the implications of the machine tools design and composition. To approximate the maximum attainable energy productivity, the impact of adjustments in performance and technical capability on the energy productivity are subsequently reflected.

4.3.3.3 Adjustment in Performance Intensity λ_E

Using the example of the grinding machine tool with the energy consumption function (4.4), the energy productivity is calculated for the performance intensity λ_E ranging from 0.5 up to 40 mm^3 s^{-1} (see Fig. 4.11). The results point out that an incremental increase in performance intensity exerts a positive influence on the energy productivity up to a material removal rate (MRR) of λ_{E^*}. This trend reverses with further advancing MRR due to the quadratic regression model of the energy consumption function. An increase in material removal rate from 4 to 8 mm^3 s^{-1} leads for instance to improvements in energy productivity ΔP_E of 76 %. With the first derivation of the functional relation for the energy productivity (4.5) approaching zero, the maximum attainable energy productivity is achieved at the performance intensity λ_{E^*}. The corresponding material removal for each invested watt-hour amounts in this case to a maximum of $P_{E^*} = 21.04\,\text{mm}^3$.

Dominating the productivity values of all other process settings, the performance intensity λ_{E^*} characterizes an energy production frontier of the machine tool (see OD' in Fig. 4.12). It correlates with the minimum attainable value of the underlying energy consumption function (4.4). The operation of the machine tool at an alternative performance intensity results consequently in inefficient activities (Sonntag and Kistner 2004). Instead of conducting a material removal at a lower performance intensity than λ_{E^*}, it is accordingly advantageous to carry out the efficient transformation and turn off the machine tool immediately after the completion of the machining task (temporal adjustment) (Albach 1962). In this case, the input-oriented technical efficiency is indicated through the distance BB' between the energy production frontier OC and the corresponding activities with adjusted performance (visualized as ABC in Fig. 4.12).

If the required removed material volume for a constrained time period exceeds the efficient material removal rate of λ_{E^*}, an adjustment in performance intensity

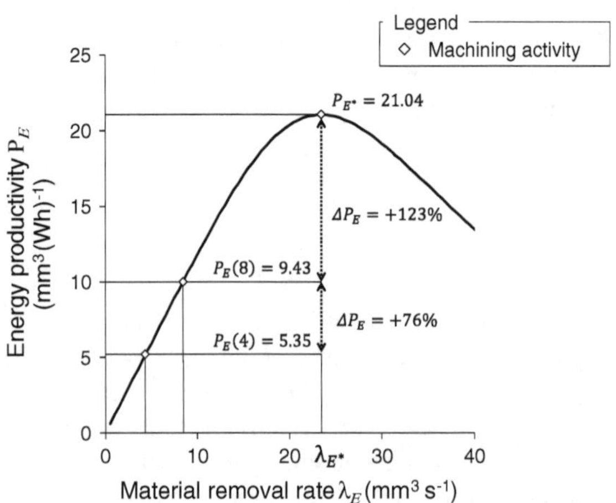

Fig. 4.11 Performance adjusted energy productivity of the grinding machine

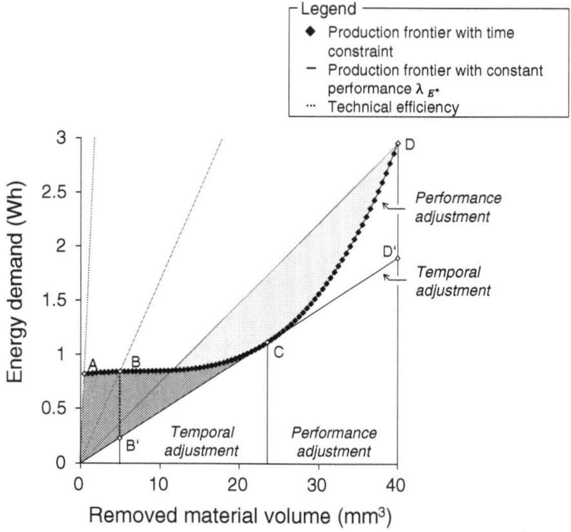

Fig. 4.12 Performance-related energy production frontier of the grinding machine tool, adapted from (Albach 1962)

is required to meet the output demands. The energy production frontier shifts correspondingly from the activities CD' to CD for performance intensities exceeding λ_{E^*} (Albach 1962).

The adjustment in time and subsequently performance intensity enables to define a process-oriented energy production frontier for the machine tool under study. Indicating process conditions with maximum energy productivity, the efficient transformations for a given energy consumption function are characterized through

- the temporal adjustment OC up to the performance intensity λ_{E^*}
- and the subsequent adjustment in performance intensity CD for a further advancing MRR (Albach 1962).

An alternative visualization of the process-oriented production frontier is provided in Fig. 4.13 depicting the efficient transformations with regard to the performance intensity.

4.3.3.4 Adjustment in Technical Capability a_E

The second influential factor on the energy productivity is the technical capability of the machine tool. It determines the specific form of the energy consumption function reflecting the energy requirements of the machine tool design to remove a defined material volume (Fandel 2010). The consumption function is bound to the characteristics of the machine tool operation and regarded as an unalterable factor (Sonntag and Kistner 2004). Adjustments in the machine tool composition create a modified technical capability requiring to reassess empirically the consumption function (Dyckhoff 2006).

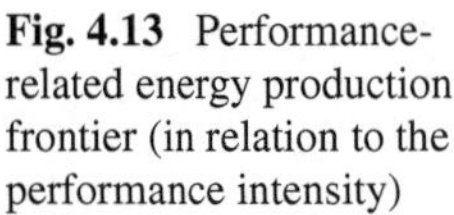

Fig. 4.13 Performance-related energy production frontier (in relation to the performance intensity)

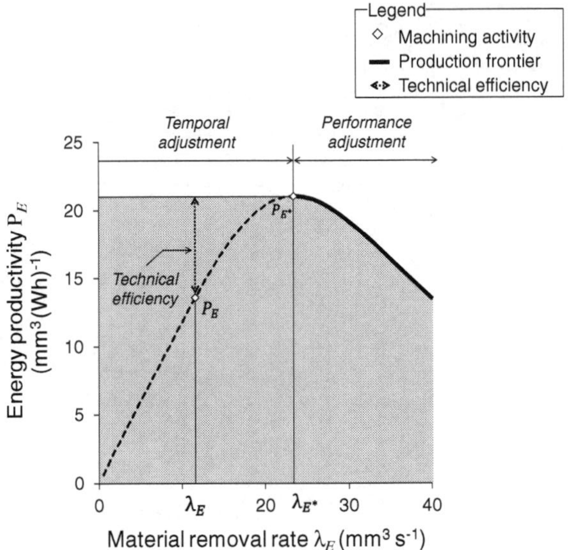

A technical capability, which maximizes the energy productivity, is characterized by a minimum achievable energy consumption function at any level of performance intensity (4.5) (Fandel 2010). Being constrained to the experimental observation, the identification of the prospected energy consumption function requires conducting comparative assessments of prevailing machine tools. In order to exemplify the assessment of a dominating technical capability, the energy consumption functions of two machine tools with the same machining technology but dissimilar technical properties are illustrated in Fig. 4.14. The consumption functions of both machine tools have an intersecting point requiring the same specific energy demand for the operation at the performance intensity λ_{EB}. For performance intensities up to λ_{EB}, a smaller specific energy demand for machine tool 2 than machine tool 1 is ascertained. This coherence reverses for performance intensities exceeding λ_{EB}.

Under the assumption of free choice between both machine tools and an identical technical feasibility, machine tool 2 exceeds machine tool 1 in energy productivity up to the performance intensity λ_{EB}. Machine tool 2 represents the dominating transformation. It outlines accordingly the efficient energy consumption function within the considered process window. As the predominance reverses with further increasing performance intensity, machine tool 1 depicts the efficient consumption function within the related operative range. Based on the individual energy consumption functions of both machine tools, a fully enclosing consumption function with minimal specific energy demands can be constructed from the dominating transformations (see Fig. 4.14). The resulting envelopment enables additionally to quantify the achievements in technical efficiency by selecting the efficient machine tools at the performance intensity λ_{EA} and λ_{EC}.

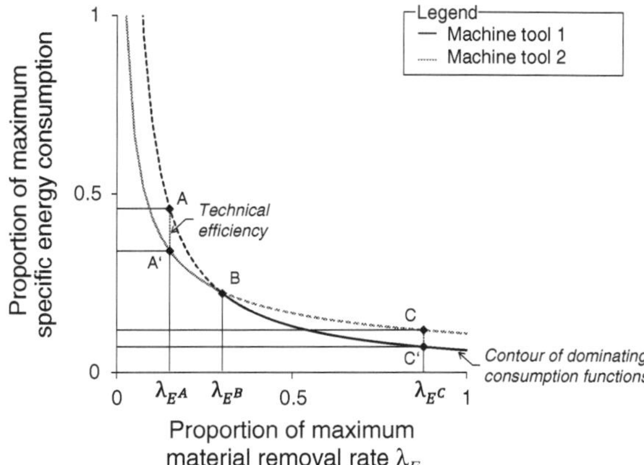

Fig. 4.14 Impact of the technical capability on the energy productivity

Representing the benchmarks in energy productivity, the involved machine tools demonstrate a superior technical capability in the respective performance sections. As a result, the efficient energy consumption function specifies an energy production frontier with adjustable technical capability. It indicates the technically achievable energy minimum of a machining technology for each level of performance.

The diversity of machine tools and operational concepts results in an extensive variety of technical capabilities. The availability of comprehensive information about the inherent energy consumption functions is furthermore limited and constrained by the time demand to carry out the measurement and perform the statistical analysis. As indicated in the state of research, the effort to characterize the energy consumption of machine tools intensified nevertheless recently providing initial parametric functions for a selection of machining processes (see Sect. 3.1.1). The results of the case-by-case analysis can therefore serve as a source of information to establish the frontier of dominating consumption functions for different metalworking technologies (see Herrmann et al. 2009; Shin 2009; Li and Kara 2011; Rajemi 2010; Diaz et al. 2010b).

The functional relations provided for instance by Kara and Li enable to compare the specific energy characteristics of five different turning machine tools. As visualized in Fig. 4.15, the Colchester Tornado A 50 represents for identical operating tasks the dominating machine tool with a minimum energy consumption function. It indicates accordingly the outline of the energy production frontier among the observed machine tools. Without a reference about the machine specific process window, the direct comparison of the energy consumption functions is based on the assumption of full validity and applicability. The anticipated assumptions can however result in misleading inferences about the energy production frontier, if the feasible process window of the dominating consumption function does not comply with the revised scope of performance. This may particularly apply for divergent process windows of manually operated as well as highly computerized machine tools and consequently limit the comparability of the energy consumption functions.

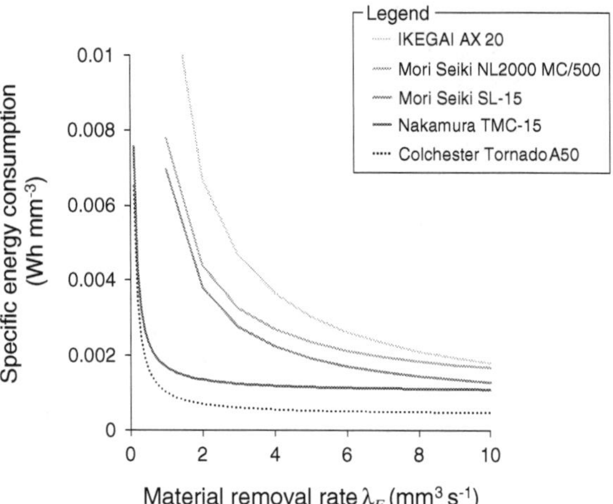

Fig. 4.15 Empirical functional characterization of the energy demand for different turning machine tools, derived from (Kara and Li 2011)

Overall, the limited availability and complex estimation obstruct the provision and comprehensive comparison of energy consumption functions to derive an efficient frontier for machining technologies. As an alternative approach, the minimum energy consumption function with adjustable technical capability is determined through data envelopment analysis (DEA). Despite the comparison of parameterized energy consumption functions, it relies as a non-parametric estimation approach initially on activity analysis establishing a discontinuous production frontier of efficient activities (see Sect. 4.3.2) (Fandel 2010; Kistner 1993).

An activity describes an operational mode of a transformation process, which is characterized through the associated input and output quantities (Koopmans 1951). Based on the assessment of activities, DEA creates a contour of linear sections gradually enveloping activities with dominating productivity (Daraio and Simar 2007; Allen and Dyckhoff 2002). The resulting envelopment provides an estimate on a production frontier for an anticipated technology. It indicates an empirical reference for the quantification of technical efficiency (Coelli et al. 2005). The form of the production frontier is determined through assumptions on the technology and properties describing the interrelations between observed activities. The assumptions emphasize predominantly the free disposability, the returns to scale and the possibility to combine activities as influencing properties of the underlying technology (Bogetoft and Otto 2011; Daraio and Simar 2007).

The first assumption on free disposability is concerned with the feasibility to dispose undesirable inputs and outputs. The production set permits accordingly activities, which create less output with more input (Coelli et al. 2005). The resulting frontier of the constructed technology is consequently defined as free disposal hull (Cooper et al. 2007). It refers exclusively to observed, dominating activities and defines the smallest enclosing boundary (Daraio and Simar 2007).

The returns to scale reflect as second assumption the ability to decrease or increase the input–output combination of an activity. While the constant returns to scale relate to the arbitrary variation of the transformation activity, the variable returns to scale revoke the possibility of scaled adjustments (Bogetoft and Otto 2011). Differentiating furthermore the capability of up- or downscaling, the assumption of non-increasing and non-decreasing returns to scale can be formulated as variations of the constant returns to scale (Allen and Dyckhoff 2002).

The third assumption contemplates the possibility to combine activities as elements of the established production set. Generally, the combination of activities can be distinguished into additive, linear and convex forms (Dyckhoff 1994). Additive combinations postulate the assumption that the summation of individual activities results in new feasible activities. Linear combinations merge furthermore the assumption of additivity with constant returns to scale. The resulting production set encloses accordingly scalable activities as well as their combinations (Allen and Dyckhoff 2002). A generalization of the linear combination is expressed with the assumption of convexity. It defines a production set that entails feasible activities and furthermore the combination of activities in any weighted average (Coelli et al. 2005). The interpolation between activities enlarges consequently the scope of the technology by defining virtual, non-observed activities (Allen and Dyckhoff 2002). The assumption of convexity enables in this way to indicate a production frontier based on the observation of few dominating activities. As a consequence, the consideration of non-observed activities is instantly linked to the risk of relying predominantly on assumed rather than empirically observed activities (Bogetoft and Otto 2011).

The specification of the assumptions affects the enveloping frontier of the technology and creates an individual, adaptive threshold referencing efficient activities (see Fig. 4.16). The characterization of the technology takes place in accordance with the scope of investigation and depends on the particular mode of perception (Allen and Dyckhoff 2002).

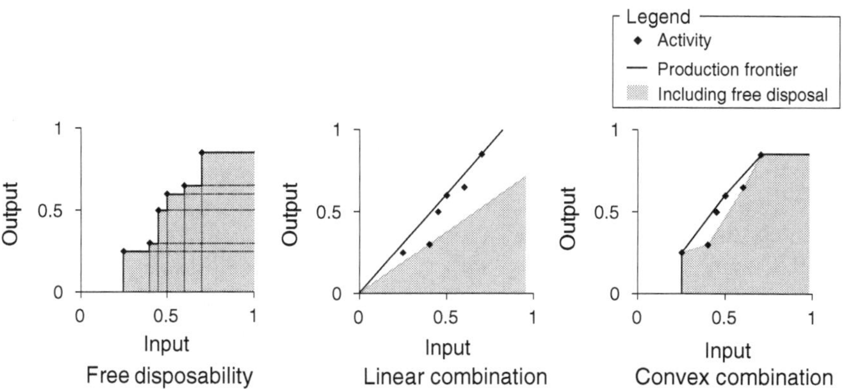

Fig. 4.16 Influence of the assumptions on the form of the enveloping frontier, adapted from (Dyckhoff 2006)

To approximate the efficient energy consumption function with adjustable technical capability, the DEA is interpreted in form of the Gutenberg production frontier. This implies to revise the definition of an efficient activity as well the assumptions on the technology within the context of consumption functions. In addition to the input–output combination, an activity can alternatively be indicated as a particular point of the consumption function (Sonntag and Kistner 2004). The activity specifies in this way an energy demand for material removal, which is associated with an unequivocal technical capability. Providing information on the related performance intensity, it designates furthermore an attainable operating condition and contributes to ensure the comparability among observed machine tools. In line with the assessment of dominating consumption functions, a dominating activity is characterized by a minimum specific energy demand at a given performance intensity.

In addition to the identification of dominating activities, the envelopment depends on the presupposition to combine and interlink observed activities with divergent technical capabilities. In contrast to the assumption of a linear technology for the introduced temporal and performance adjustment, the technological properties of adjustments in the technical capability are not integrated in the Gutenberg production frontier (Sonntag and Kistner 2004; Kistner 1993). The technical capability is fundamentally considered as a constant factor defining a unique consumption function (Dyckhoff 2006). The form of the consumption function depends on the underlying characteristic of the machine tool, which is established with the completion of the design and development process (Sonntag and Kistner 2004). Adjustments in technical capability can therefore be assumed possible throughout the design phase of the machine tool. Being able to adapt and merge the superior technical capabilities of dominating machine tools, the possibility to design a new machine tool with shared technical properties corresponds with the assumption of convexity. As the proposed concept for energy performance management is concerned with the design as well as the use phase of machine tools, the convex combination is presumed as technological property for the approximation of the enveloping frontier.

Using the DEA to estimate the efficient energy consumption function with adjustable technical capability, the frontier is constructed based on dominating linear interpolations of efficient activities. These are dominated by neither alternative activities nor interpolations between other efficient activities. Integrating the assumption on free disposability, the envelopment frontier considers additionally activities from all feasible consumption functions with higher specific energy demands. An example of a DEA frontier for consumption functions, which satisfies the assumption on convexity and free disposability, is illustrated in Fig. 4.17. It visualizes the estimation of the theoretical production frontier through the approximated efficient energy consumption function with adjustable technical capability. The resulting envelopment defines accordingly the technically achievable minimum specific energy demand for a machining technology for each considered performance level. It indicates an empirical reference of superior technical capabilities, which maximizes the energy productivity.

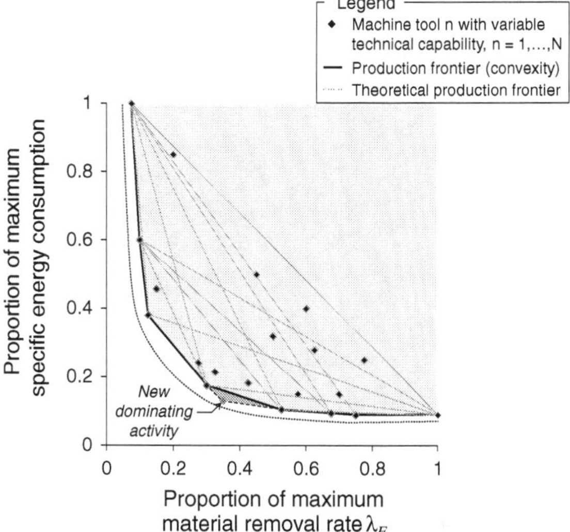

Fig. 4.17 Approximation of the energy production frontier for machining technologies

The estimation of the technically achievable consumption function with minimal energy demand through DEA provides three particular advantages. First, the activity analysis demands solely to observe single process conditions and derive the associated specific energy demand and material removal rate, instead of conducting a series of experiments for statistical analysis. This reduces the risk of failure due to a time-consuming data acquisition and facilitates furthermore to create a comprehensive dataset with input provided by machine tool manufacturers as well as users. Secondly, the DEA eases the continuous refinement of the envelopment adapting the frontier to newly added dominating activities (see Fig. 4.17). Starting with an initial set of activities, the approximation of the minimal energy can be elaborated in detail and expanded to include wider range of performance intensity. Thirdly, the orientation of activities provides first indications on possible improvements through adjustments of the influencing variables energy demand, material volume and material removal rate. With regard to the steep gradient of the energy consumption function for small performance intensities, the adjustment in performance intensity provides for instance an immediate benefit enhancing the energy productivity without major modifications of the removed material volume or energy demand of the machine tool.

Based on the determination of the efficient consumption function, the technical efficiency of non-dominating activities is indicated through the distance between the activity and the corresponding linear section of the enveloping frontier. Figure 4.18 visualizes the assessment of inefficiency for the activity Z, which is compared to the activity Z' as efficient reference with identical performance intensity. The activitiy Z' represents a virtual activity of the linear function interlinking the observed dominating activities A and B. With an increasing number of dominating activities, the dependence on virtual activities for reference decreases enclosing the frontier steadily with observed activities for direct comparison (Kistner 1993).

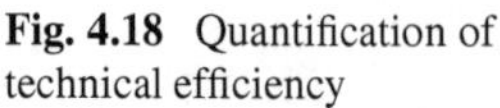

Fig. 4.18 Quantification of technical efficiency

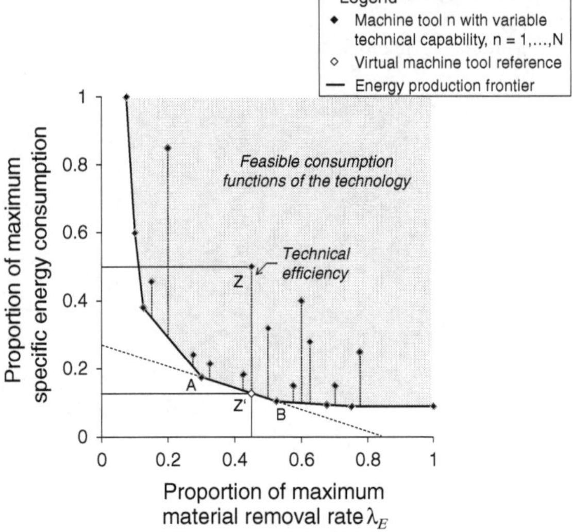

Characterizing the energy usage in machine tools in form of the Gutenberg production frontier, the energy production frontier represents a methodological approach to specify technically achievable minimum energy demands with particular consideration for the process and machine tool design. With regard to the technical capability of a machine tool, the empirically observed reference defines a threshold of best available solutions for a machining technology. It designates consequently a marginal process energy gradient and does therefore serve as a means to define target values for energy performance measurement.

4.3.4 Formulation of Energy Performance Limits for Machine Tools

The formulation of performance limits is a vital element triggering the process of performance measurement. With the specification of technically achievable energy values as performance limits for machine tools, improvement towards the designated threshold is encouraged. Concluding the results of the preceding analysis, a set of three measurable limits for the energy performance of machine tools is deduced (see Fig. 4.19). While the sphere of influence of the first two relates to the machine tool design, the efficient process design obtains a process-oriented perspective.

Adopting the systems-based perception of an ideal energy demand, the first performance limit is concerned with the process-independent, fixed power consumption of machine tools. To avoid the energy overhead in non-operating times, a target value of zero is stipulated as performance limit for the fixed power consumption during idle modes.

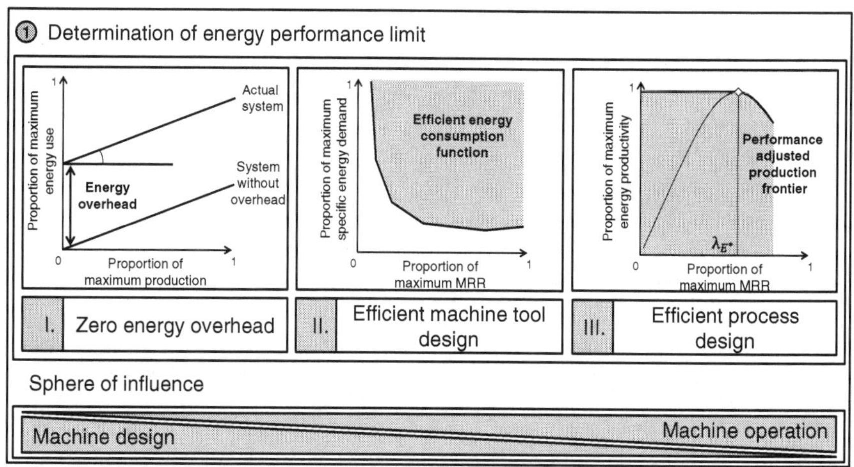

Fig. 4.19 Formulation of energy performance limits for machine tools

The second performance limit specifies a marginal process energy demand with regard to the machine tool design based on the efficient energy consumption function of the underlying machining technology. For defined output demands and related process requirements, the machine-oriented energy production frontier provides empirical guidance on the technically achievable minimum energy demand.

The third performance limit maintains a strong scope on the selection of process parameters indicating a performance intensity for the operation of existing machine tools with a maximum energy productivity (or correspondingly minimum specific energy demand). Based on the individual, fixed energy consumption function of the machine tool under study, a minimum energy demand is determined with regard to adjustments in the process design.

The determination of quantitative metrics for the second and third performance limit presupposes the empirical estimation of the energy production frontiers for the machining technology and the investigated machine tool. While the characterization of a single consumption function expresses an individual scope, the approximation of efficient energy consumption functions for machining technologies demands to revise a variety of different machine tools. A first insight into the energy requirements of single manufacturing activities is for instance provided by (Gutowski et al. 2006) collecting published data in literature (see Fig. 4.20).

The motivation to generate comprehensive, contemporary datasets can be strengthened especially through collaborative efforts. Recent initiatives reviewing the energy demands of various machining technologies are for instance mentioned in Pohselt (2011) and Kellens et al. (2011a). These can provide initial datasets for the approximation of efficient energy consumption functions for machining technologies. With regard to the negligence to release internal corporate data, it has to be pointed out that the conversion into normalized activities ensures that no utilisable suppositions about the related original application can be extracted. The formulation

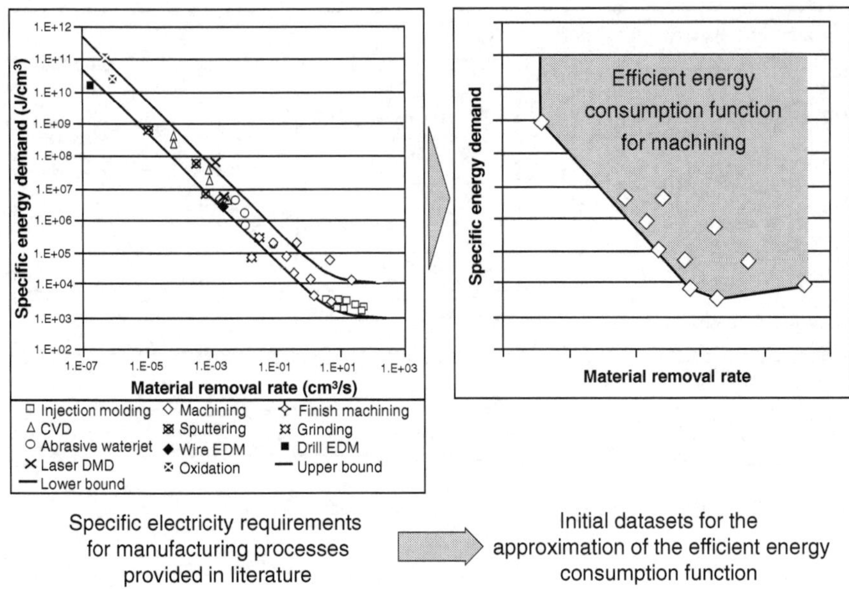

Fig. 4.20 Specific electricity requirements for a selection of manufacturing processes, adapted from (Gutowski et al. 2006)

of energy limits based on activity analysis is accordingly open to consider datasets from a variety of information sources empowering the involvement of machine tool builders as well as users.

4.4 Measurement of Actual Performance for Machine Tools

In addition to the formulation of energy target values, the quantification of the energy efficiency gap necessitates the determination of the actual energy performance for an investigated machine tool. It expresses an obtained level of the parameter value on the indicator scaling. To determine the parameters for a machine tool through measurement, methodological means are required to ensure a reliable and valid acquisition of input data.

4.4.1 Measurands of the Energy Performance Indicators

The structural form of the performance indicators delineates the measurands of the energy limits, which have to be revised in order to ascertain the performance for a machine tool under study (see Fig. 4.21). The first energy limit relates to

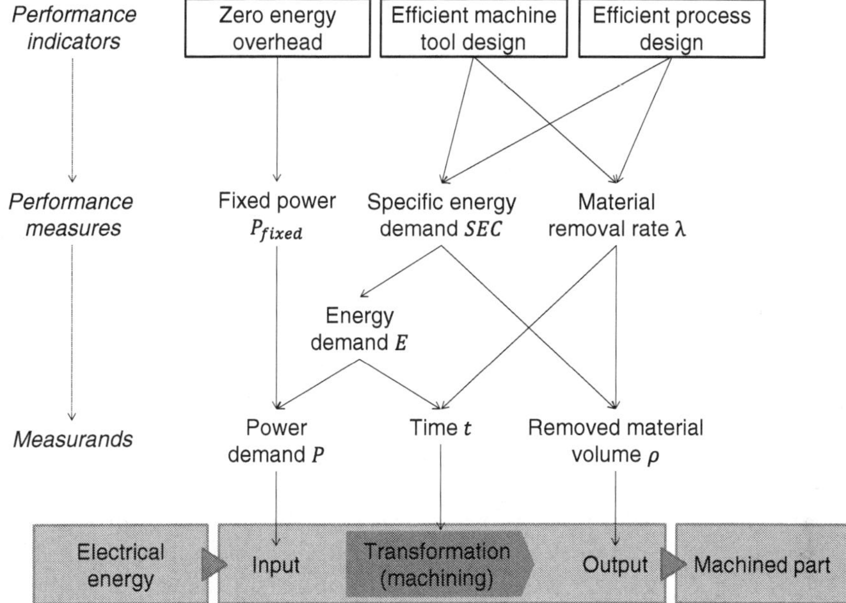

Fig. 4.21 Required measurands to determine the energy performance indicators

the energy overhead with the fixed power demand as an absolute performance measure. The complementary energy limits characterizing the efficient machine tool and process design refer to a machining activity with the measured variables power demand, time and removed material volume.

The measurands time and material volume (or the combination of both in form of the material removal rate) represent commonly established quantities, which are estimated throughout the product and process design using for instance computer aided planning systems (Eversheim and Schuh 2005; Kalpakjian and Schmid 2001). In case this information is not available, the measurands can alternatively be determined through time recording and weighing of parts. To obtain data about the individual power consumption, it is however imperative to conduct measurements on investigated machine tools using electrical measurement equipment. The installation and operation of a measuring device is a particular aspect of a planning process to conduct experimental power measurements. It is specified in the following section in detail.

4.4.2 Planning of Experimental Power Measurement

Measurement contemplates the determination of reliable and valid quantities through the experimental estimation of parameter values. The process of measurement involves generally the realization of a *measuring method* and *procedure* as well as the *aggregation of measured values* (see Fig. 4.22) (Fridman 2012).

The measuring method is concerned with the specification of the actual sensing technique and properties to estimate the parameter (Fridman 2012). Using the Hall-effect principle, the power consumption of machine tools is obtained through the measurement of voltage and current in three-phase systems (Dyer 2001). Kara et al. 2011 revise a set of market available mobile and permanently installed measuring devices delineating common technical properties. The examined devices incorporate the sensing process and signal transformation providing the power consumption as output data. Taking the distortion in phase between current and voltage into consideration, the resulting parameter values can be differentiated into the apparent, reactive and effective power demand (Parthier 2010). All examined measuring devices are equipped with the ability to perform dynamic measurements allocating output values commonly at a temporal resolution of 1s. The outliers among the market available equipment surpass this resolution up to a minimal temporal interval of 0.015s. The highly resolved illustration of power dynamics is instantly challenging the data capturing and analysis (Herrmann et al. 2010).

In addition to the preparation of the metering device, the measuring procedure provides formal guidance on the execution of operations during the measurement. It comprehends accordingly the definition of measurement conditions and processes (Fridman 2012). Before conducting measurements on machine tools, it is important to note that the power demand is influenced by the process-induced operation of the energy consumers within a machine system (see Sect. 2.2.1). In addition to the machine inherent components (e.g. drives), the revised system

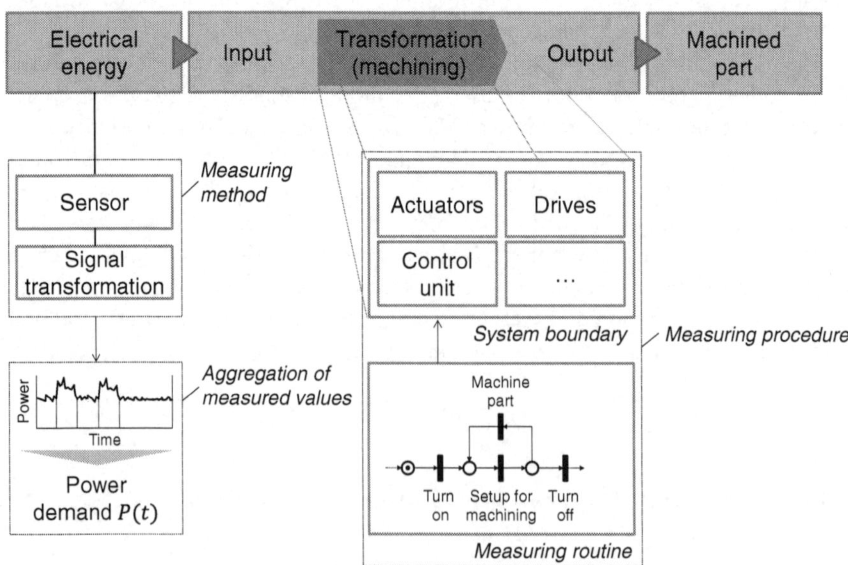

Fig. 4.22 Process elements of the experimental power measurement

boundary may include energy-consuming supplements as handling devices, supplementary filters, mist collectors or external hydraulic units. In order to maintain consistency among different measurements, the measuring procedure for machine tools is accordingly challenged to define a unified system boundary as well as a measuring routine with uniform operational machining tasks.

Striving to guide the process of power measurement through procedural approaches, the working group of the International Standards Organization currently elaborates testing routines for the upcoming ISO 14955 series on the Environmental evaluation of machine tools (Hagemann 2011). In addition, the Japanese Standards Association pursues an alternative approach proposing a reference workpiece to assess machining procedures (Japanese Standards Association 2010). An analogous test piece is introduced by Behrendt et al. (see Fig. 4.23). It is specifically designed to determine and compare the power requirements for milling machine tools describing different machining operations and process sequences (Behrendt et al. 2012).

The conceptual design of test pieces for a standardized power measurement is generally complex with regard to the intended machining tasks and modes. Furthermore, test pieces are yet predominantly defined for milling and drilling operations. The approaches go far beyond the necessary measurement requirements to determine the energy performance indicators. The compulsory operational modes, which have to be measured for the actual performance metrics, are exemplified in Fig. 4.24. These involve the *waiting for parts* to derive the fixed power demand as well as the *material processing* to deduce the specific energy demand of an activity for the efficient machine tool and process design.

In line with the assignment of operational modes, the measuring routine necessitates furthermore to identify the distinctive machine modes within the power characteristics in order to ensure the traceability of the measurement. The challenge of recognizing operational conditions is visualized for a grinding machine

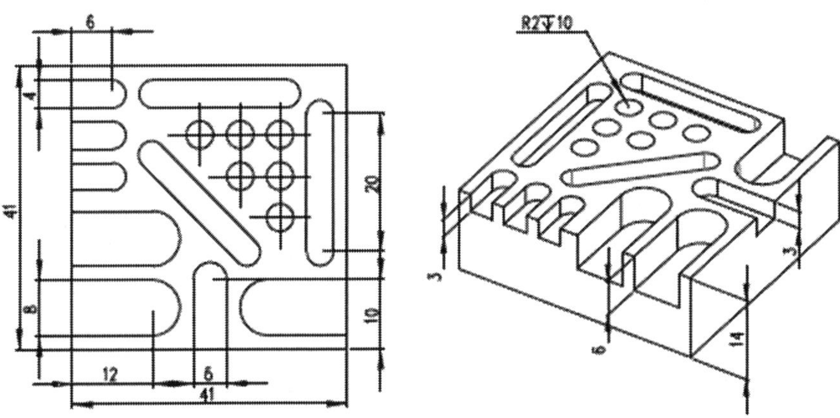

Fig. 4.23 Standard test piece (small scaled) (Behrendt et al. 2012)

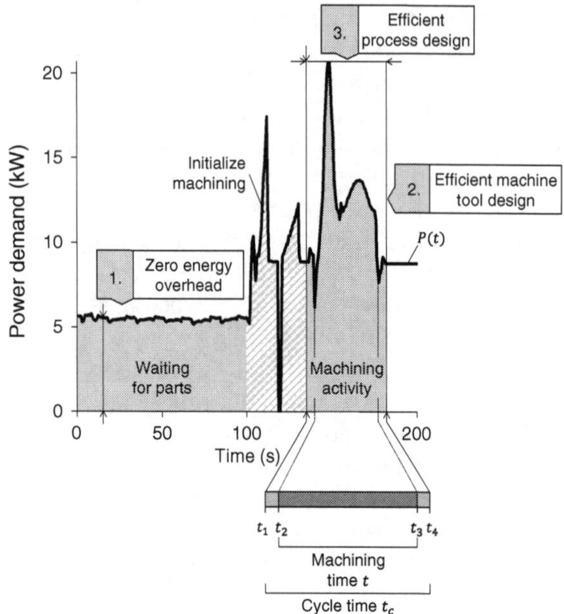

Fig. 4.24 Compulsory operational modes within the measuring routine

tool in Fig. 4.25. The measured profile illustrates the power characteristics of two machining cycles, which are interrupted in between due to missing parts. As the underlying operational modes cannot be distinguished directly and unequivocally, the measuring routine relies on a formalized set of timed operations involving the waiting for parts and the material processing. It can for instance be composed of a predefined temporal period in waiting mode followed by a recurring process sequence with a fixed number of parts.

The final element of the measurement process is concerned with the aggregation of measured values. The complex machining dynamics within machine tool

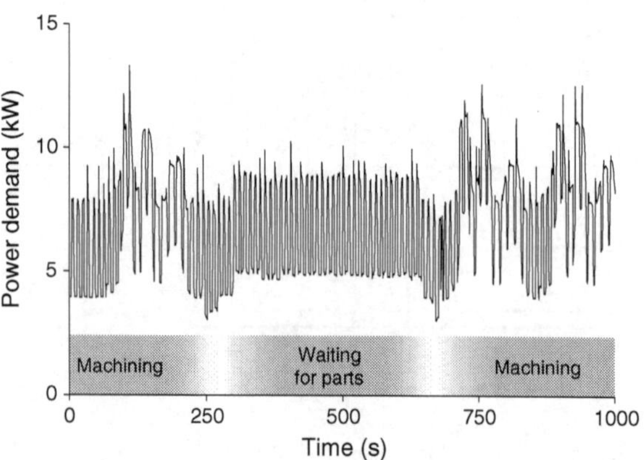

Fig. 4.25 Distinction of operational machine modes during power measurement

systems as well as the accuracy of the power measurement can produce divergent measurement results for analogous measuring procedures. In order to minimize the influence of disturbances and errors on the measured values, the statistical averaging for the parameter aggregation is emphasized compensating the deviation through repetition. With the introduction of the measuring procedure, an immanent opportunity is instantaneously established to balance the influence of statistical discrepancies on the aggregation of measured quantities.

4.4.3 Definition of a Measuring Strategy for Machine Tools

Formalizing all elements of the preceding measurement process, an unequivocal measurement strategy for machine tools is defined to quantify the power demand in experiments. It provides guidance for the determination of the actual performance based on a uniform and consistent power measurement for a machining activity. Bearing in mind the origins of the organizational barrier, the practicability of the measurement approach is established under the premise to minimize the complexity, effort, time and associated costs (Pfeifer and Schmitt 2010). The measuring strategy to determine the actual energy performance of a machine tool is defined as follows:

4.4.3.1 Measuring Method

The specification of the measuring device relates to the minimum requirements and elementary functionalities, which have to be considered for the selection and configuration of equipment. With regard to the fundamental capabilities of measuring devices, an extensive specification of minimum requirements is however negligible. The measured quantity value to determine the actual energy performance measure is the effective power demand indicating the useful work (Capehart et al. 2012). For the power measurement, a temporal output resolution of 1 s is defined. It facilitates to observe the dynamics of the material removal and operational behaviour of the entire machine tool system balancing the level of detail and associated complexity of data handling (see Fig. 4.7) (Kellens et al. 2011a; Herrmann et al. 2010).

4.4.3.2 Measuring Procedure

Reviewing the definition of a unit process in Sect. 2.2.1, the system boundary encloses all electrical energy consumers within a machine tool system, which provide an essential and immediate contribution to perform the value creation through material removal. The measuring procedure relates furthermore to the use phase of machine tools with realistic, application-oriented machining activities. The measuring routine is composed of two elements (see Fig. 4.26). The first element designates a period of several minutes in waiting mode maintaining a steady power demand. The second element follows immediately with a sequence of ten identical machining tasks. These are processed consecutively. Monitoring both elements in a single measurement routine requires accordingly a minimal time demand. It exerts also only a minor influence on the prevailing machining schedule in industrial applications.

4.4.3.3 Aggregation of Measured Values

To minimize the impact of stochastic deviations, the first energy metric regarding the energy overhead is estimated as the averaged fixed power demand within the observed waiting time. The specific energy demand for the second and third energy performance indicator is calculated by integrating the effective power demand over the time. Averaging the calculated energy demand for each of the ten machining sequences ensures to compensate random variations in the processing and measurement conditions.

As indicated in Fig. 4.24, the time demand of the power integral can be referenced to either the machining or the cycle time. While the machining period relates to the time demand to perform the processing (as productive use), the cycle time includes additionally the setup and idle times (Rogalski 2011). In the interest of operability, the cycle time simplifies the observation of the actual performance. The energy demand can be estimated as an equal share of the integral of power over the entire machining sequence. However, the estimation enfolds in this calculation also the power demand in non-productive times, which can be influenced by external factors (e.g. the conveyer or handling system). In order to maintain a distinctive scope on the machine tool system, the machining time is accordingly referred to as temporal basis for the assessment of the actual performance. For that reason, the energy demand is determined based on the power consumption within the machining time.

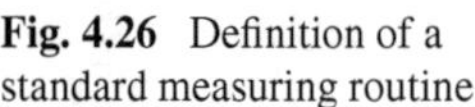

Fig. 4.26 Definition of a standard measuring routine

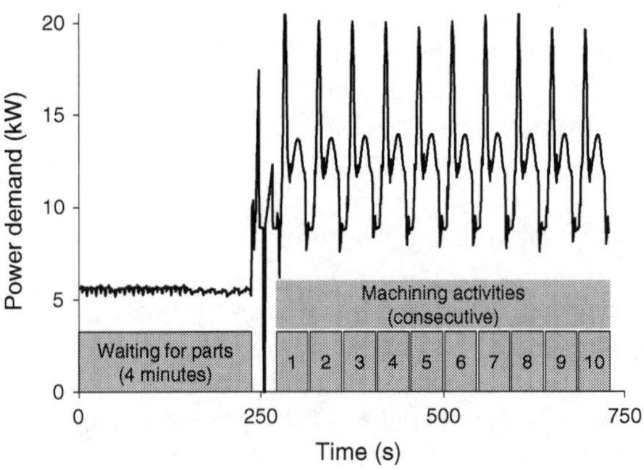

With the formulation of a unified measurement strategy, the second module of the concept for energy performance management describes an approach to determine the actual performance of machine tools in a consistent and reliable way. The formal aspects of the measurement strategy are summarized in Table 4.2. Standardizing the conversion of measured quantities into performance indicator values, the general applicability as well as traceability and effectiveness of the experimental power measurement is ensured. Both concept modules for the evaluation of energy efficiency provide together the essential information for the quantification of the energy efficiency gap for machine tools. Completing the process of performance measurement, the identified technical efficiencies represent the essential input to initiate the energy performance management.

4.5 Energy Breakdown Analysis

4.5.1 Specification of Potential for Improvement

The quantification of the energy efficiency gap designates an attainable improvement potential for an actual machine tool in relation to a corresponding energy performance limit. The energy performance measurement assigns the technical efficiency from a systems perspective to the entire machine tool challenging the identification of improvement opportunities. Only the third energy performance limit regarding the efficient process design advises instantly means to enhance the performance (see Fig. 4.13). These comprise the adjustment in time and material removal rate. In

Table 4.2 Specification of the standardized measurement strategy

Measurement process	Element of the measurement strategy	Specification
Measuring method	Measured quantity	Effective power demand
	Temporal output resolution	1 s
Measuring procedure	System boundary	All electrical energy consumers with an essential and immediate contribution to perform the material removal
	Measuring routine	4 min waiting time for parts with steady power demand 10 consecutive machining tasks
Aggregation of measured values	Performance metric on the energy overhead	Mean power value within the observed waiting period
	Performance metric on the efficient machine and process design	Mean energy demand of the machining sequences calculated from the integral of power over the machining time

contrast, the efficiency gaps for the first and second energy performance criteria concentrate on the divergence in the power demand ΔP_{fixed} and energy requirements ΔE as visualized in Fig. 4.27. The reasons for the inefficiency within complex machine tool systems remain however unspecified, which obstructs the immediate assignment of improvement means and the realization of prospective gains in efficiency.

The actual energy performance for both indicators originates in the power consumption of the machine tool elements in waiting mode and during the machining period. The factors influencing the individual power demands are heterogeneous and emerge indirectly in the structural composition of the machine tool as well as directly in the degree of automation and the energy conversion of the component (Wolfram 1986; Binding 1988).

In order to conclude means to improve the actual machine tool structure (e.g. integrate lightweight materials), indications on the inefficiency can be obtained through a comparative assessment with the physical design features of the ideal system reference. As the identified means affect the fundamental composition of the machine tool, the implementation remains however commonly restrained to elementary modernizations (e.g. retrofitting) or subsequent machine tool generations (Zein et al. 2011). The automation and energy requirements of electrical components enclose alternatively instant opportunities to enhance the technical efficiency. These comprise for example the adaption of operational routines as well as the substitution with energy-saving alternatives (Wolfram 1986; Zein et al. 2011). As a prerequisite for the identification and exploitation of component-based improvement opportunities, it is necessary to gain awareness about the operation and power demands of the energy-related equipment. To ease the analysis of the diverse machine tool elements, the first concept module on improving the energy efficiency supports the identification of relevant components, which contribute substantially to the actual energy performance.

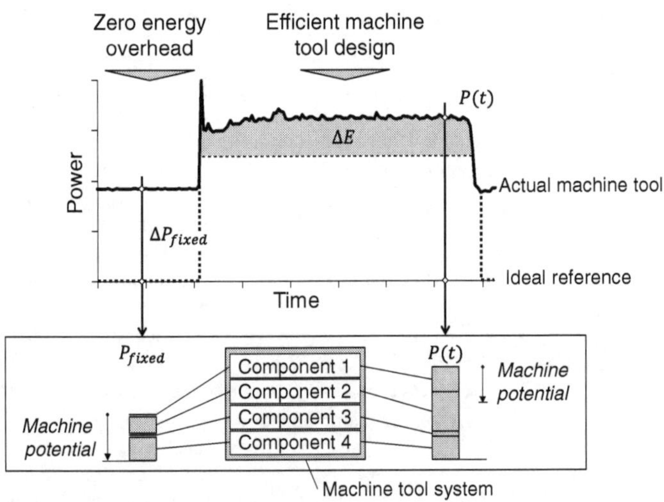

Fig. 4.27 Origins of the actual energy performance in machine tool systems

4.5.2 Concurrent Power and Machine Data Analysis

While the predictive simulation is yet limited in scope (see Sect. 3.1.1), the experimental measurement is the only feasible alternative to disclose data about the operation and associated power demands of the machine tool elements. In order to extract the necessary data with high accuracy, the installation and operation of power measuring equipment is required at each electrical component. As a result, the isolated observation is confronted with extensive effort, time demand and costs. This impedes a practical application.

Striving to minimize the effort for power measurement, a screening approach is established with the aim of constraining the detailed assessment to prioritized system elements. It estimates the contribution of a component to the actual energy performance based on the rated power and the temporal operation in waiting mode and during machining. In accordance to the power specification for machine tools (see Sect. 2.2.2), the rated power provides as an attribute of electrical devices an approximation on the potential capacity. It is usually documented in the wiring scheme and component manuals (Li et al. 2011). In contrast to the input-oriented perspective for machine tools, the power rating for electrical equipment can also be associated to the mechanical output power (as applied for motors) (Müller 2001). In order to compare individual components, it is therefore important to maintain a consistent perspective on the power rating. To convert an output rating into the corresponding electrical input capacity, predefined energy conversion ratios for components can be used (e.g. electrical motor efficiency) (Abele et al. 2011; Thumann 2010).

In order to obtain the additionally required information for the screening approach, a monitoring system is established to track the operational behaviour of the electrical elements. It extracts process signals from the machine tool control unit, which indicate the temporal activation of the integrated components, in analogy to a control-oriented condition monitoring (Plapper and Weck 2001; Brecher and Weck 2006). Extending the scope towards a concurrent power and machine data analysis, the experimental monitoring provides furthermore the opportunity to anticipate the associated power demands of energy-using devices (Herrmann et al. 2011a). In addition to the priorization of components, this can eliminate the need for otherwise necessary assessments on component level. The conceptual outline of the screening approach is visualized in Fig. 4.28 differentiating the three elements data acquisition, data analysis and priorization.

4.5.2.1 Data Acquisition

The data acquisition is concerned as the first element with setting up the power measurement and establishing the connection to the control system of the machine tool. In accordance to the measuring method described in Sect. 4.4.3, the effective power demand of the entire system is obtained using an electrical measuring

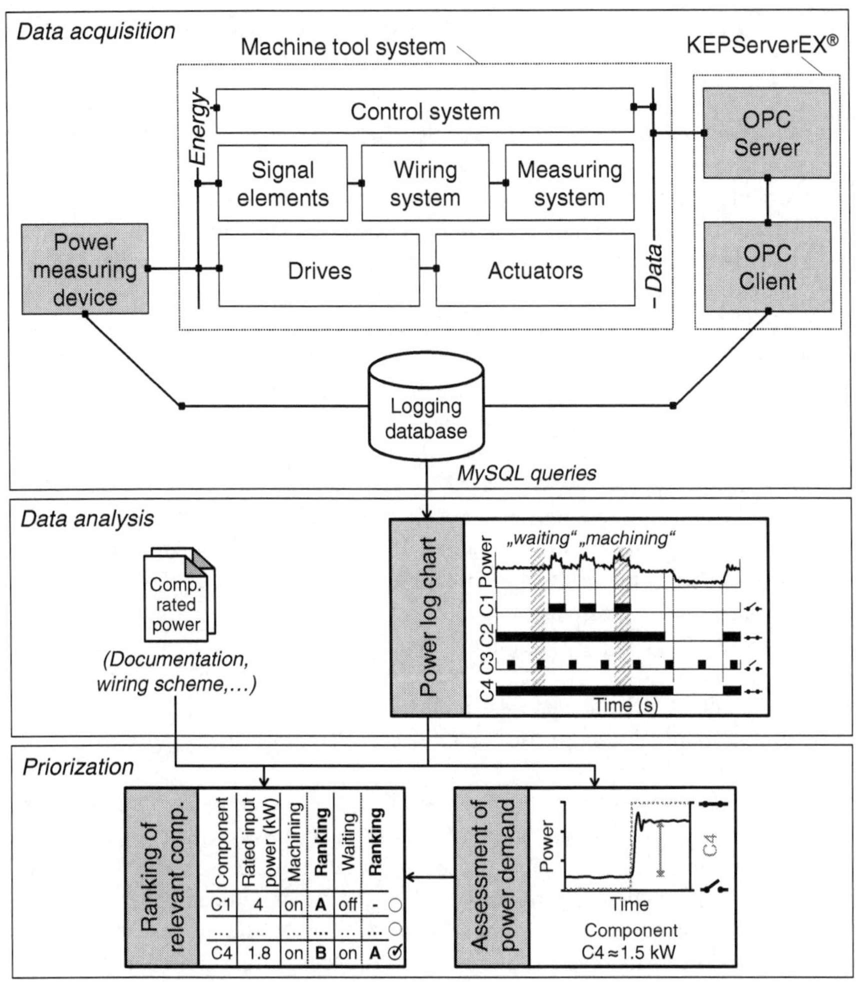

Fig. 4.28 Concept elements for the concurrent power and machine data analysis

device with a temporal resolution of one second. The applied device is equipped with the functionality to transfer the measured quantities continuously to a logging database. Alternatively, a temporal series of measured data can be stored in the device and included in the database after the measurement.

The data connection to the machine control is realized through an Ethernet network and software applications as KEPServerEX® using standard communication mechanisms as OLE for process control (OPC) (Herrmann et al. 2011a). The OPC application interlinks the logging database with the programmable logic controller (PLC) as part of the control system. Based on the analysis of input signals from the operator or internal measuring systems, the PLC controls the operation of the individual electromechanical processes within a machine tool system (Tönshoff

1995; Weck and Brecher 2006a). It represents an important source of information reflecting the actual mode of the components in form of dynamic process signals (Brecher and Weck 2006). The assignment of the signals to the integrated devices is documented in the wiring scheme facilitating the identification of equipment within the investigated machine tool (Deutsches Institut für Normung e.V. 2006).

In addition to the primary interest in the operation of components, the extracted data can also enfold process signals, which track for instance the productive and non-productive mode of the entire system (e.g. in process). The measurement of machine data can accordingly provide a beneficial input for energy-related time studies and analysis (Vijayaraghavan and Dornfeld 2010). Apart from the monitoring functionality, the minimum requirements for the measurement routine include essentially the observation of the waiting and machining mode as defined in Table 4.2. The extracted data from both sources is logged in a joint database coupling the power demand and process signals of components based on identical time tags. It is therefore indispensable to maintain a consistent time reference throughout the measurement.

4.5.2.2 Data Analysis

The second element of the experimental approach encompasses the activities to revise the measured data using power log charts and to compile the input power rating for components. With regard to the power rating, this involves especially the collection of component-specific data from the machine documentation and the conversion of rated output values into the input power demands.

In line with the charts used for machine condition monitoring, a power log chart visualizes synchronously the dynamic power consumption and alteration of process signals during a sequence of machining operations (Brecher and Weck 2006). It provides the basis to recognize the active elements of the machine tool system in waiting mode and during the machining of parts as input for the priorization. The impact of components on the power consumption becomes furthermore apparent clarifying the overall dynamics throughout the measurement. Identifying couplings between the power consumption and the activation of components, the individual power requirements of machine tool elements can subsequently be approximated under certain conditions.

In Fig. 4.29, a log chart for an exemplary grinding machine tool illustrates the results of a concurrent power and machine data analysis. It demonstrates the process signals of seven identified elements from the wiring scheme of the examined machine tool. The log chart omits the process signals of two axis drives for translational movements, which are part of the drive system and not specifically listed among the publicly available PLC signals.

The analysis of the data reveals that the power demand in waiting mode (W) is connected to the operation of the spindle cooling, hydraulic and control unit. These three components are continuously activated and temporarily complemented by the operation of the tool and workpiece spindles during the machining process (M). Among the

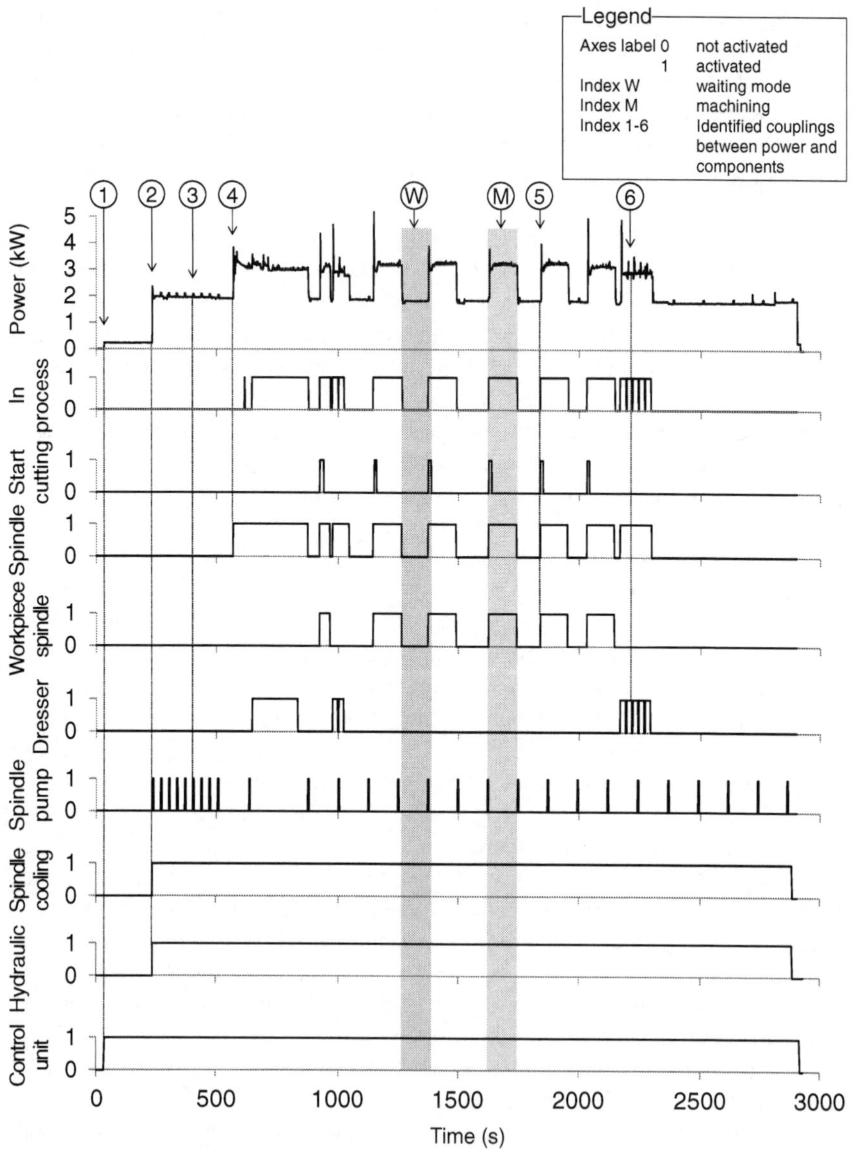

Fig. 4.29 Power log chart for an exemplary grinding machine tool

remaining components, especially the periodical activation of the spindle pump has to be pointed out. It operates only for certain proportions of the revised periods. Based on the collected data, six couplings between the power demand and the activation of components are disclosed throughout the monitored sequence (see 1–6 in Fig. 4.29). These are specified in the final element of the concurrent power and machine data analysis.

4.5.2.3 Priorization

Within the priorization part of thc screening approach, the gathered information on the power ratings and operational behaviour is combined in order to approximate the influence of each revised component on the actual energy performance. It is determined as the product of the normalized rated input power and the proportional activation during machining and in waiting mode. Sorted in descending order, the resulting quantities delineate a priorization of machine elements for both observed modes. Concentrating on the components with a substantial influence on the energy performance, the ranking provides guidance for the detailed power measurement and eases the review process for improvement opportunities within machine tool systems.

A ranking of components is demonstrated in Table 4.3 for the studied grinding machine tool. The power ratings are extracted from the machine documentation and normalized to the maximum value. The output-oriented power ratings for the spindles are transposed into an input demand assuming a conversion ratio of 80 % for spindle units (as depicted by Abele et al. 2011). The proportional operation during machining and in waiting mode is obtained from the power log chart in Fig. 4.29. The resulting order of the involved components indicates a substantial impact of the spindle during machining and the spindle cooling unit in waiting mode. The effort to investigate improvement potential in detail should therefore focus at first on these components.

The priorization of components valuates the potential contribution to the actual energy performance for subsequent analysis. However, it does not describe the genuine power demand. By extending the measuring scope from the process signals towards the concurrent analysis of power and machine data, the impact of

Table 4.3 Ranking of relevant components for detailed energy breakdown analysis

Component (rated input power)	Normalized rated power (%)	Active in machining (%)	Priorization "machining" (%)	Active in waiting mode (%)	Priorization "waiting" (%)
Spindle (13.75 kW)	100	100	A (100)	0	–
Workpiece spindle (included in spindle rating)	100	100	A (100)	0	–
Dresser (2.75 kW)	20	0	–	0	–
Spindle pump (0.11 kW)	0.8	10	E (0.1)	0	–
Spindle cooling (1.1 kW)	8.0	100	B (8.0)	100	A (8.0)
Hydraulic (0.09 kW)	0.7	100	D (0.7)	100	C (0.7)
Control unit (0.3 kW)	2.2	100	C (2.2)	100	B (2.2)

activating components becomes apparent in the power dynamics of the log chart (see Fig. 4.29). Allocating an induced increase in the overall power demand to the originating component, the power consumption of an element within the machine tool system can accordingly be specified as an averaged growth in power throughout the activation period. Applying this approach to the exemplary grinding machine tool, the mean power consumption of integrated components is quantified and visualized in Table 4.4. It includes the power demand of the high-prioritized spindle as well as the control unit, spindle pump and dresser. As a result, the conjoint analysis can substantially reduce the effort to conduct detailed power measurements leaving the spindle cooling unit, workpiece spindle and hydraulic to be observed.

The decision to assign a change in the power demand to a component is primarily challenged to ensure that the operation of other devices is not compromising the measured quantities. The impact of activating the hydraulic and spindle cooling unit is for instance coupled and therefore connected to an unequivocal increase in the power demand during the start-up procedure of the machine tool. The individual contribution of each device is accordingly not clearly distinguishable. As an alternative example, the activation of the workpiece spindle does not seem to affect the overall power requirements (see Fig. 4.29). In situations, in which the impact of a component on the power demand is undetectable or superposed by other activated components, the immediate extraction of information about the power consumption is accordingly restrained. Nevertheless, the analysis reveals a stronger influence of the hydraulic and spindle cooling unit on the power demand than the workpiece spindle. These results advance in this way the planning of detailed power assessments redirecting the effort from the workpiece spindle with an unobserved impact directly towards the spindle cooling unit as a direct successor in priority.

4.5.3 Proposition of Improvement Opportunities

The conversion of the technical efficiency for the first and second energy performance criterion into improvement means presupposes information about the individual operation and power demands of the electrical elements within an investigated machine tool system. The concept module on the energy breakdown analysis establishes with the generic screening approach a simple and quick priorization of relevant components. It provides a basis for the identification and exploration of improvement potentials.

With regard to the proposition of improvement opportunities, the concurrent power and machine data analysis represents with the power log chart an extensive source of information. It creates transparency about the overall power dynamics in relation to the operating behaviour of the integrated components. As a consequence, the outcome of the conjoint analysis enables directly to revise potential for improving the energy performance through adjustments in the temporal operation

Table 4.4 Approximation of power demands to components

Point of observation	Measured power and machine data	Approximation of power demands
①		Control unit $P_C = 0.25$ kW
②		Hydraulic/spindle cooling unit $P_{HS} = 1.6$ kW
③		Spindle pump $P_{SP} = 0.1$ kW
④		Spindle $P_S = 1.2$ kW

(continued)

Table 4.4 (continued)

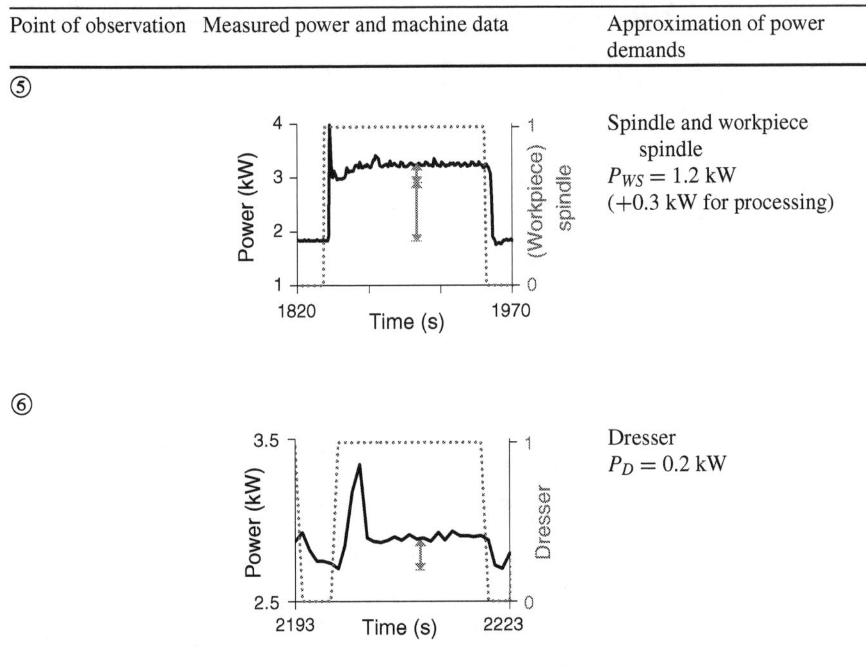

Point of observation	Measured power and machine data	Approximation of power demands
⑤		Spindle and workpiece spindle $P_{WS} = 1.2$ kW (+0.3 kW for processing)
⑥		Dresser $P_D = 0.2$ kW

Axes label: 0, not activated; 1, activated

of the electrical devices. This applies especially for the activated equipment in waiting mode resulting in a demand for control-oriented improvement measures to disconnect the identified elements. By gaining additional insight into the individual power demands, the energy breakdown analysis designates a reference for the design and evaluation of enhanced solutions. Concerning the second energy performance criterion, the review of means for the relevant components encompasses accordingly the enhancement of the power consumption as well as the adaption of the temporal operation.

Revising the power log chart and component priorization for the exemplary grinding machine tool, the results enable to conclude starting points for the improvement of the power demand in waiting mode and during machining. These concentrate especially on the deactivation of the spindle cooling unit, control unit and hydraulic in non-operating times to approach the criterion of the zero energy overhead. With regard to the second energy performance criterion, the effort to enhance the energy performance is specifically concerned with the exploration of improvement potentials to reduce the power demand of the spindle and spindle cooling unit. While the operation of the spindle is indispensable to pursue the machining process, the cooling of the spindle unit may also be sufficient with discontinuous work cycles (as implemented for the spindle pump). In contrast to control-oriented means, efforts to reduce the power demand are usually associated with a substitution of components. This challenges the realization of improvement within established machine tool systems due to technical and economic constraints.

4.6 Improvement Planning

4.6.1 Interlinking of Energy Performance Limits and Improvement Measures

The realization of improvement in the actual performance of machine tools necessitates the specification and implementation of adequate measures for each energy limit. In contrast to the third energy performance criterion, the energy breakdown analysis constitutes an interim step to deduce potential saving opportunities from a component perspective for the first and second energy performance criterion. Utilizing these results, the final concept module of the energy performance management introduces with the improvement planning a mapping scheme, which supports the assignment of measures to exploit these potentials (Zein et al. 2011). As visualized in Fig. 4.30, it establishes an interconnection between the functional requirements expressed through the energy efficiency gaps and appropriate measures.

The technical efficiency associated with the first energy performance criterion relates to the power consumption of activated components in waiting mode (exemplified as C1–C4 in Fig. 4.30). This power draw can for instance be avoided by turning off the corresponding components. In order to accomplish this functional requirement and realize the designated potential, the improvement planning has to focus accordingly on the identification and assignment of measures, which target the deactivation of components.

The second performance limit is concerned with the enhancement of the energy usage in the machine tool system expressing a functional need for design options to reduce the energy demand and conversion losses. Revising the factors, which influence the energy demand, various opportunities and associated improvement measures can be identified. These concentrate on the reduction of time and power demands as well as the reuse and recovery of energy losses (Zein et al. 2011). As a result, the improvement planning is challenged to review and specify distinct means from the manifold possibilities.

As an alternative, the performance limit regarding the efficient process design stipulates inherently distinct improvement measures. Based on the energy consumption function of the investigated machine tool, it includes either the temporal adjustment operating at the maximum energy productivity or the variation in performance intensity to meet the required output demand. Contrary to the yet unspecified possibilities to deactivate components and reduce the energy requirements, the elaboration of definite means to realize the efficient process design is therefore accomplished.

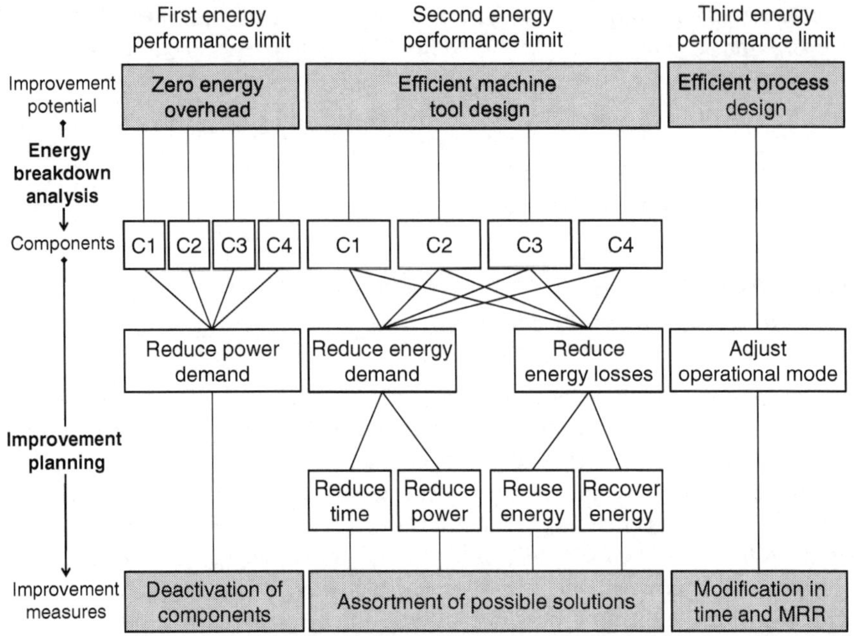

Fig. 4.30 Mapping of improvement potential to measures

4.6.2 Specification of Improvement Measures

Technical and organizational measures to enhance the usage of electrical energy in machine tools are diverse and continuously extended in availability through innovative developments and refinements of existing technologies (e.g. improved energy conversion in motors) (see Sect. 2.2.3). The subsequently revised specification of distinct improvement measures for the first and second performance limit intends to provide an insight into currently available measures in order to guide the improvement of the actual energy performance.

To approach the performance limit of a zero energy overhead, the deactivation of components in waiting mode obtains a primarily control-oriented perspective. In the simplest form, the machine tool elements are turned off by the operator manually or by supplementary devices (e.g. DMG Energy Save) in non-operating times (Li et al. 2011; DMG 2012). The implementation of these measures is instantly appropriate for machine tools involved in a discontinuous job or small batch production with substantial waiting times between machining orders. This includes also disconnecting the machine tool after the last working shift and on weekends.

For shorter non-operating periods in higher scaled production environments, the deactivation can be realized through automated, control-integrated shutdown routines. An exemplary concept for the deactivation of components within machine tools is documented by Gleason. It encompasses five consecutive steps to disconnect individual elements from the power line (Gleason 2011). As visualized in Fig. 4.31, the approach involves turning off components at the end of certain periods Δt_X in waiting mode. After disabling the machine light, supplementary systems as well as basic machine tool elements are gradually deactivated. While the concept description does not delineate the prospective power reduction and time intervals, it does however provide an indication on the general feasibility for a variety of machine tool elements. In designing the measure, it is recommended to turn off the identified energy-relevant components at an early stage. Furthermore, the reactivation time Δt_{RX} should be kept at a minimum level in order to avoid the depletion of obtained energy savings while returning to a state of operational readiness. The implementation demands consequently an individual adaptation and planning of the measure for the revised machine tool system.

Extending the scope from the control of binary operating states towards multiple energy-enhanced modes, adaptive turn-off strategies for each element of the machine tool system can be realized (Siemens 2010). The combination of devices with hibernating modes and an energy-aware control concept are currently proposed as central features of PROFIenergy, which is established as a supplementary module for automation systems (PROFIBUS Nutzerorganisation e. V. 2011).

Compared with the specification of solutions for the distinct functional task to deactivate components, the selection of measures to enhance the energy demand of a machining cycle is of increasing complexity. This is mainly due to the four possible objectives, which can be pursued by implementing technical and organizational approaches, as well as the requirements and underlying effects of mechanisms and strategies (Zein et al. 2011). This obstructs the definition of generally

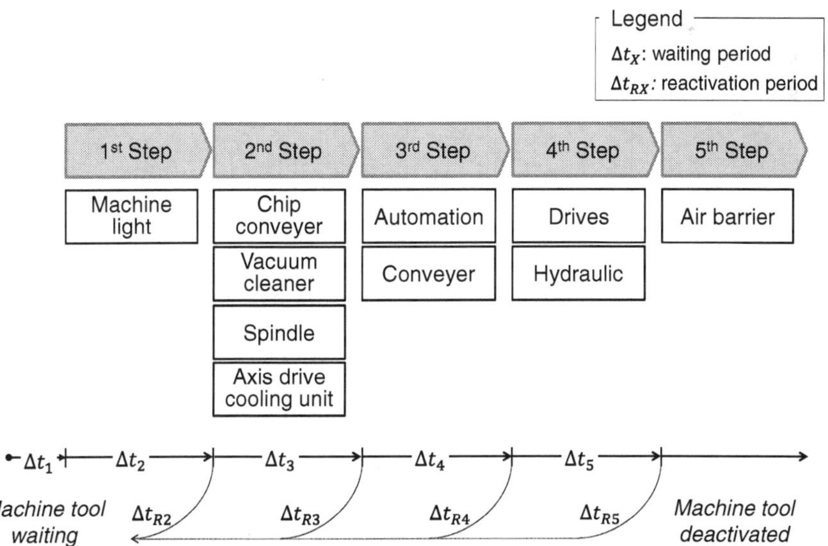

Fig. 4.31 Example of a deactivation routine for components, based on (Gleason 2011)

admitted and valid improvement approaches for machine tools and demands to revise means from an increasing diversity of options for the individual elements and entire machine tool system.

For that reason, a generic routine is established based on the envisaged objectives of measures to support the identification and derivation of distinct approaches (see Fig. 4.32). It implies a preferential order to initially reduce the time and power demand in advance to reduce the energy losses. This structure ensures that the opportunity to deactivate a component is for instance favoured over the replacement with an energy-enhanced substitute. In a subsequent step, the feasibility to apply potential measures, which contribute to the objective, is reviewed. With regard to the reduction of power demands, this encloses for instance activities to adjust, dimension or substitute components. The detailed elaboration of measures requires predominantly specific knowledge about the requirements and operating conditions of components within the machine tool system. Accordingly, the application of design and planning tools may be necessary, for example, as the *Siemens Sizer* for the software-based dimensioning of drive systems (Siemens 2011).

In line with the systems approach for energy improvements described by Stasinopoulos et al. the routine considers at first the entire machine tool system revising opportunities to reduce the time demand within the machining cycle. Secondly, the specification of measures is conducted on a component level following the defined ranking of the energy breakdown analysis.

To support the review of measures for each defined objective, an indicative overview is compiled in Table 4.5. It does not raise any claim to completeness as new technical and organizational solutions continue to emerge. The list of potential measures aims particularly at supporting the review process triggering the identification of distinct means for improvement (Binding 1988). Accordingly, it encompasses general, application-independent approaches to enhance the actual energy demand of a machine tool system and its elements throughout a machining cycle.

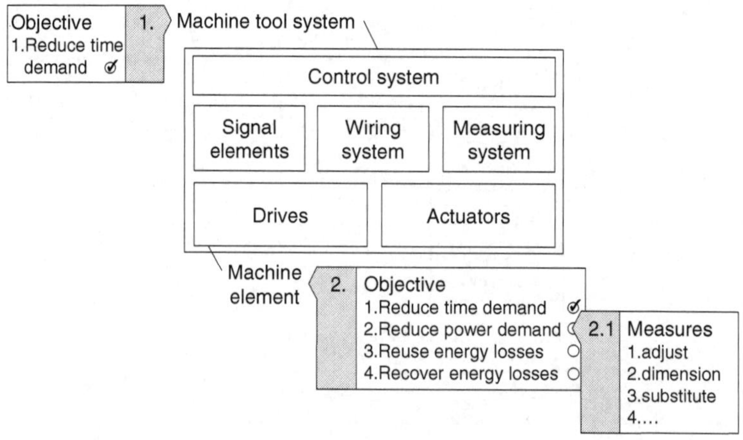

Fig. 4.32 Routine for the specification of improvement measures

Table 4.5 Overview of measures to enhance the energy demand for machining, based on (Binding 1988; Zein et al. 2011; Schischke et al. 2012b)

Reducc time demand
Minimize machining allowance (offset)
Minimize removed material volume
Reduce process sequences (e.g. energy-enhanced NC-programmes)
Minimize traverse distances
Reduce activation time of components
Operate components only when necessary
…
Reduce power demand
Dimension components to suit the appropriate load
Reduce moved masses
Apply components with enhanced energy conversion ratios (e.g. energy labels)
Minimize friction, damping, electric losses,…
Substitute component with energy-enhanced alternative (e.g. converter instead of transformer)
…
Reuse energy
Reuse mechanical energy
Conserve kinetic energy
Transform energy into other useful forms of energy
…
Recover energy
Recuperate heat losses through heat exchanger
Conserve energy losses to main thermal stability of the machine tool
…

While the routine does not relieve the necessity to design and plan distinct means to enhance the technical efficiency of an investigated machine tool, it does provide a consistent routine to contrast opportunities in form of functional requirements to measures. Together with the characterization of solutions for the deactivation of components, both conceptual approaches provide a substantial contribution to assign and elaborate distinct design options in order to realize improvement in the actual energy performance of machine tools.

4.6.3 Formulation of Improvement Policies

The preceding activities to interlink improvement opportunities with definite measures for each energy performance limit result in a specific assortment of possible means. Based on the identified technical and organizational solutions, an improvement policy is subsequently formulated, which defines a sequential order for the implementation.

From a simplified economic perspective, a fundamental precondition for the consideration of a measure within an improvement policy is the ability to ensure a general profitability at least in the long term (Gahrmann et al. 1993). A measure can commonly be considered as profitable if the revenue compensates the costs. In addition to this first ordering criterion, a ranking can furthermore be established by organising the measures in descending economic profitability (Gahrmann et al. 1993).

Necessary input data for the economic valuation represent the costs to implement the measure and the monetarily assessed saving potential. The attainable potential is determined based on the perceived reduction in time or power consumption compared to the initial situation prior to the implementation of the measure. By referencing the energy conservation for example to a period of one year, the annual improvement potential can be specified for the economic evaluation. Using the example of deactivating a machine tool in waiting mode, the saved energy requirements depend accordingly on the attained power reduction as well as the frequency and magnitude of non-operating times. As a consequence, the quantification of an energy saving potential necessitates a critical review of the actual operational behaviour in order to designate a realistic saving potential for the assessment of profitability.

The static comparison of revenue and costs facilitates the formulation of an improvement policy. Nevertheless, it has to be pointed out that this valuation does not consider the amount and timing of scheduled future cash flows. The determined profitability does therefore only represent a rough and quick estimate (Schmid 2004). Treating energy improvement measures as an investment in a technical system, the prevailing dynamic methods of investment analysis should be applied evaluating the profitability in accordance with the capital value method (Ostertag et al. 2000).

With the formulation of an improvement policy, the fourth module of the concept for energy performance management concludes the identification and planning of means to enhance the actual performance of machine tools. Together with the energy breakdown analysis, both concept modules provide essential methods and tools to convert the technical efficiency into improvement opportunities for complex machine tool systems and deduce specific means to bridge the energy efficiency gap. Completing the process modules of performance management, the energy performance measurement is initiated again evaluating the effect of implemented means on the technical efficiency.

4.7 Integration of Concept Modules into a Continuous Improvement Process

4.7.1 Workflow for Energy Performance Management

The developed concept modules of the preceding sections represent elements of the energy performance management for machine tools. In order to guide the application of the involved methods and tools, the four concept modules are implemented in a workflow model. It characterizes the processes to quantify the energy efficiency gap for machine tools and support the improvement towards an ideal energy limit.

The continuous improvement process of performance management originates from the *Plan, Do, Check, Act (PDCA) cycle* as an elementary method of systems engineering (see Sect. 2.3.3). The elaboration of the underlying problem-solving cycle with the phases measuring, learning, planning and implementing is described for performance management in diverse workflow models, which differ in the considered phases and stages (Bourne et al. 2003; Olson 2006). In accordance with the fundamental dimensions of performance management, the workflow model to bridge the energy efficiency gap for machine tools is composed of two process models targeting the evaluation and improvement. The first process element focusses accordingly on the measurement of performance including the activities to define a goal and scope, to initialize the required energy limits as well as to quantify the energy efficiency gap between the actual and ideal performance. The second process element concentrates on the management of means to enhance the energy performance and realize improvement. As visualized in Fig. 4.33, the activities in the process models form a continuous improvement cycle to enhance the energy performance of machine tools.

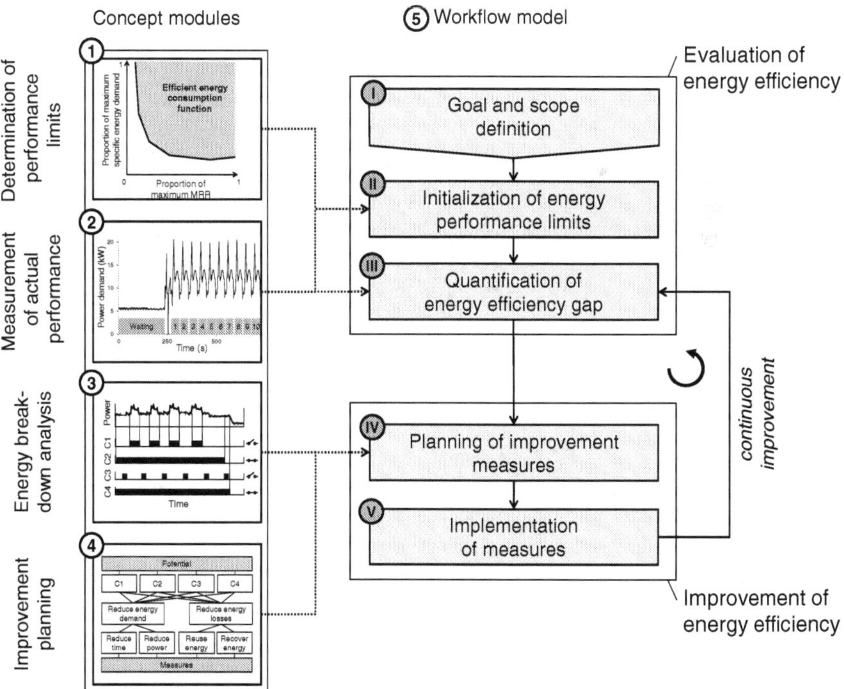

Fig. 4.33 Workflow for energy performance management for machine tools

The subsequent sections describe the operational routines compiling the methods and tools of the concept modules. The incorporated activities indicate a generic application of the overall concept, which can be adjusted with regard to the particular goal and scope (within the energy planning and implementation of industrial energy management systems).

4.7.2 Evaluation of Energy Efficiency

The first activity to evaluate the energy efficiency encompasses the *goal and scope definition* (*I*) for the energy performance management. It includes the specification of the objective and scope for the continuous improvement as well as the review of required indicators. This implies also to assign a machining technology and outline a consistent technological system boundary to ensure the comparability among the revised machine tools and processes.

The *initialization of energy performance limits* (*II*) concentrates on the determination of the ideal reference values. While the first performance indicator regarding the energy overhead represents a static value of zero, the defined activities focus on the quantification of the technically minimum achievable values for the efficient machine tool and process design (see Sect. 4.3.4). The workflow for the definition of the energy limits is visualized in Fig. 4.34.

As the formulation of the performance limit relies on an empirical observation, the process to derive the efficient machine tool design starts with the definition of an assortment of entities, which represent the investigated machining technology. In order to balance the impact of an individual machine tool on the frontier of the efficient consumption function, an amount of at least ten different machine tools with varying removal rates is initially assumed. The power consumption of each system is subsequently revised using the standardized measurement strategy introduced in Sect. 4.4.3. The strategy appoints no further regulations regarding the process parameters or used parts for the machining task. The observed material removal should nevertheless be set in accordance with commonly used process windows or preferably reflect a range of industry applied operations.

Based on the aggregated specific energy demand and associated material removal rate for each revised system, the efficient machine tools are identified. These represent machining activities, which dominate the energy requirements of alternative systems operating at the same material removal rate as well as linear interpolations of other efficient machine tools. The dominating linear functions, which interlink efficient machining tasks, delineate ultimately the efficient energy consumption function of the revised technology (see Fig. 4.17). The approximation of the enclosing frontier is only necessary once in order to initiate the evaluation of performance. It is then continuously updated and increased in detail throughout the following concept application.

In contrast to the determination of the enveloping functions for a technology, the third performance limit is concerned with the characterization of the energy

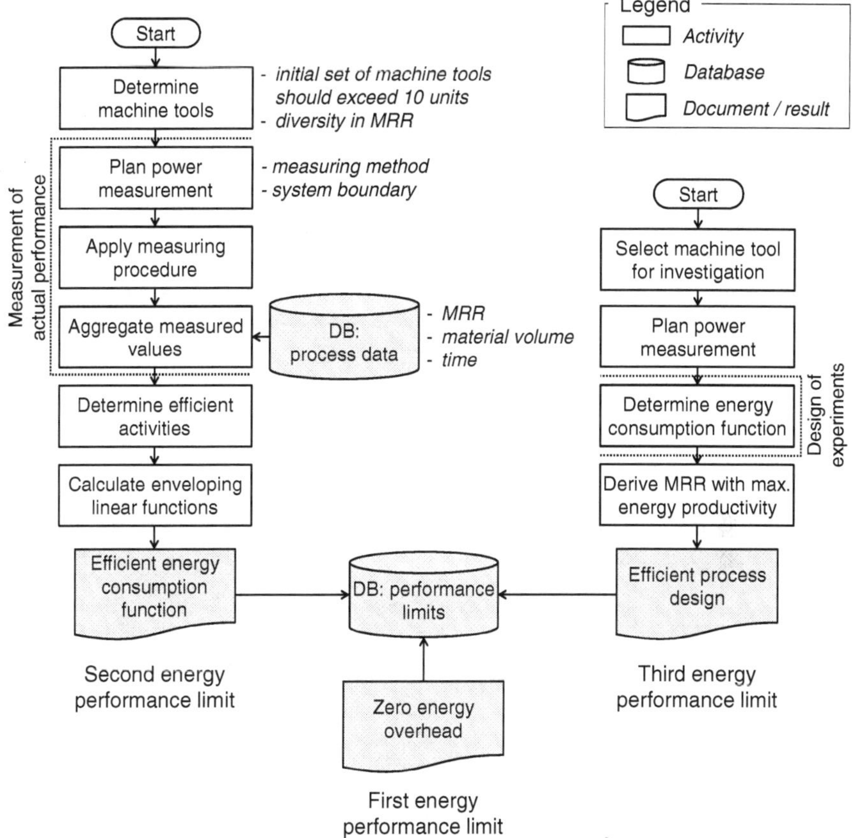

Fig. 4.34 Workflow to initialize energy performance limits

consumption function for a single machine tool. Instead of following the standardized power measurement, it necessitates a characterization of the energy requirements using design of experiments (see Fig. 4.8). This implies to conduct a series of machining tests with material removal rates (MRR) covering the machine process window (Herrmann et al. 2009; Kara and Li 2011). An example of this process is provided in Sect. 4.3.3. Based on the deduced functional relation, the MRR with the highest attainable energy productivity (defined as removed material volume per energy unit) represents the targeted indicator value.

The determination of the performance limits for machine tool and process design constitutes the first and indispensable step for the evaluation of energy efficiency. In order to cope with the necessity to assess various machine tools, collaborative initiatives (e.g. CO_2PE Cooperative Effort on Process Emissions in Manufacturing) with multiple resources can substantially ease the formulation of comprehensive and contemporary consumption functions for technologies.

The values for the energy performance indicators are input data for the *quantification of the energy efficiency gaps* (*III*) in the third step of the process model. While the energy limits for the zero energy overhead and efficient machine tool design represent comparative units for any machine tool of an examined technology, the third energy limit is constrained to the original system. Accordingly, two courses of activities are introduced describing the evaluation of the actual and ideal energy performance. The routine for a machining technology is visualized in Fig. 4.35.

The quantification of the efficiency gap for the first and second energy limit is triggered by a request to investigate the energy performance a machine tool. The examined entity is assessed using the standardized measurement strategy to specify the fixed power demand in waiting mode and the energy demand of a

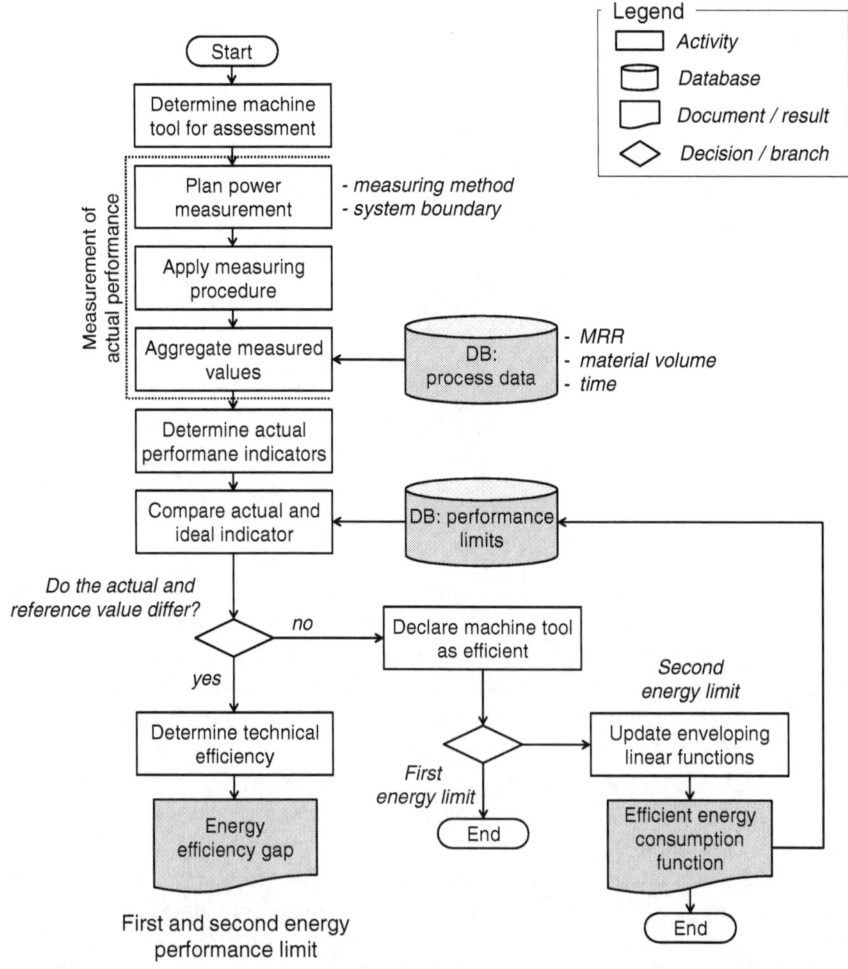

Fig. 4.35 Workflow to quantify the efficiency gap for the first and second energy performance limit

machining activity at a particular material removal rate (see Sect. 4.4.3). Both values are then related to the corresponding energy limits taken from a performance databasc. If the actual energy performance underscores the related threshold value, the divergence in the indicator units represents the technical efficiency for the actual machine tool. The identified improvement potential of the revised system is documented as energy efficiency gap. However, if the actual energy performance of the system is equal to or better than the current reference, it is assigned as a new efficient entity. For the second performance limit, this machine tool is then incorporated as part of the efficient consumption function for the technology. With the adjustment and update of the enveloping linear functions for further reference, the evaluation process for the efficient machine tool is dismissed.

The assessment of the energy efficiency gap for the third energy performance limit builds upon the preceding characterization of the related energy consumption function (see Fig. 4.36). The originally applied MRR of a current machine tool operation is for this purpose compared to the corresponding removal rate inducing the maximum attainable energy productivity. As visualized in Fig. 4.13, the divergence in the energy productivity of the actual and ideal process enables to quantify the efficiency gap. The identified potential is provided as input to the improvement process model. If the energy productivity of the actual and ideal machining parameters matches, the process is consequently declared as efficient. This terminates the evaluation process.

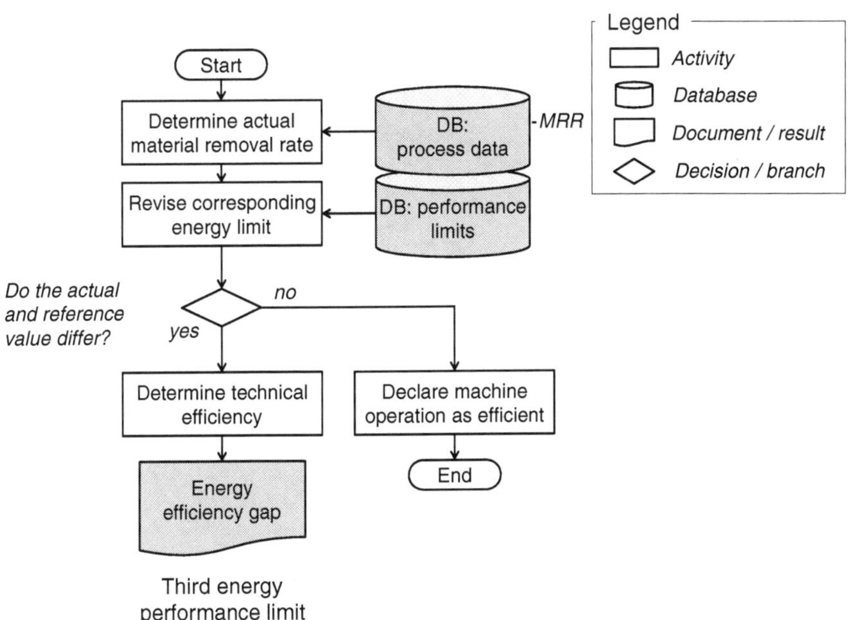

Fig. 4.36 Workflow to specify the energy efficiency gap for the third performance limit

4.7.3 Improvement of Energy Efficiency

Upon completion of the evaluation process model, the identified energy efficiency gaps trigger the management activities to identify and assign means for the enhancement of the actual performance towards the ideal reference. In line with the two evaluation routines, the second process model maintains the differentiation of actions from the machine tool and process perspective.

The activities to *plan the improvement measures* (*IV*) for the first and second energy limit begin with the energy breakdown analysis. It is an interim step to guide the allocation of saving opportunities to the electrical components (see Sect. 4.5). As illustrated in Fig. 4.37, it encompasses the concurrent measurement

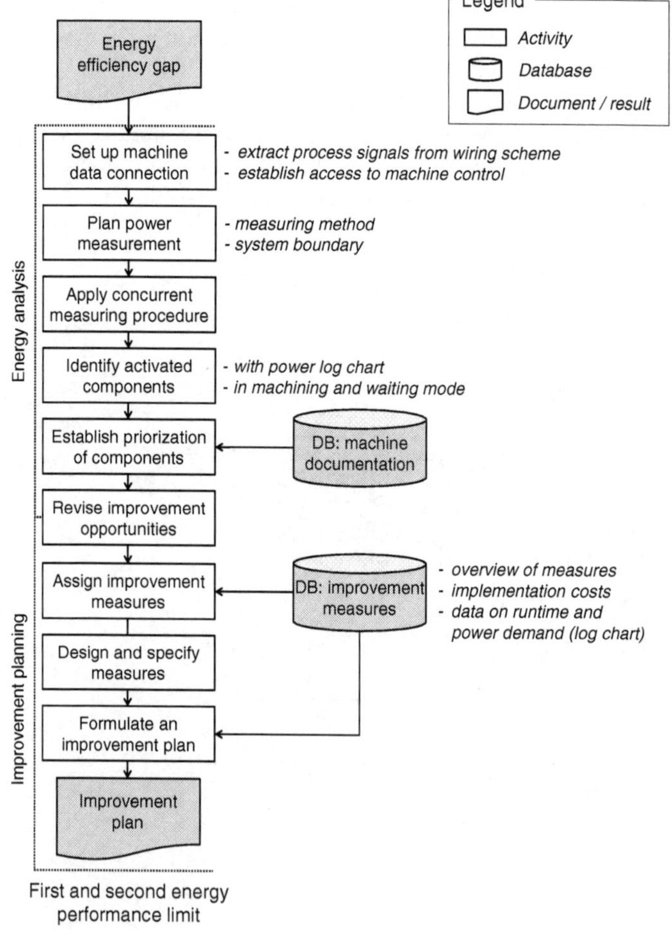

Fig. 4.37 Workflow to plan improvement towards the zero energy overhead (first energy limit) and efficient machine tool design (second energy limit)

of machine control data and power consumption. Monitoring the operation of the machine tool elements in waiting mode and during machining, the simultaneous measurement is used to identify the active components. Interlinking the obtained results about the temporal operation with the rated input power demand of the electrical entities, an energy-oriented priorization of components is established. This ranking defines an ordered structure for the review of saving potentials and assignment of measures. Within the improvement planning, an interlinking of potentials and measures is established based on a generic catalogue of design options (see Fig. 4.32).

Specifying the technical requirements for the implementation of the measures and evaluating the resulting energy saving potentials, an order of profitable means can be deduced with methods of investment analysis. The list of measures is concluded in an individual improvement plan for the first and second energy performance limit to prepare the decision about the implementation.

Due to the direct dependence of the energy efficiency gap on the material removal rate, it is possible to avoid the activities of the energy breakdown analysis for the third energy performance limit. The workflow relates instead directly to the planning of improvement (see Fig. 4.38). It encompasses the adjustment either in time or performance intensity. With regard to the variation in the process parameters, it is furthermore essential to ensure the machine and process capability approving the applicability of the aspired processing conditions. With the estimation of the saving potential for the selected measure, the formulation of the third improvement plan is completed.

The improvement planning provides three sets of measures for the energy performance limits. These focus generally on the deactivation of power demands,

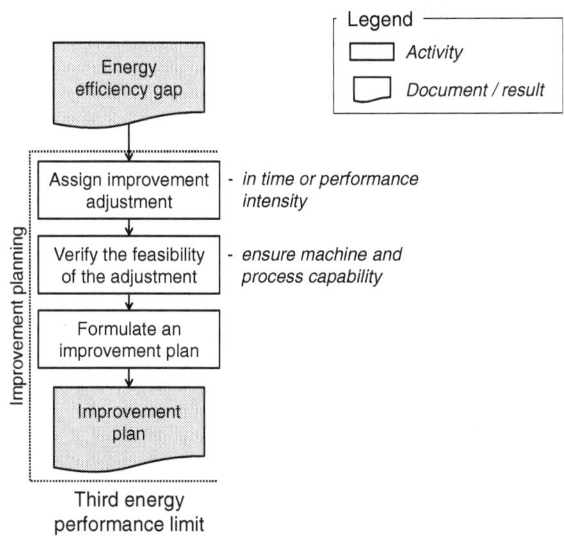

Fig. 4.38 Workflow to specify improvement for the efficient process design (third energy limit)

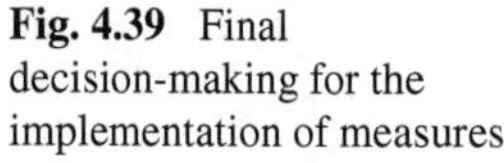

Fig. 4.39 Final decision-making for the implementation of measures

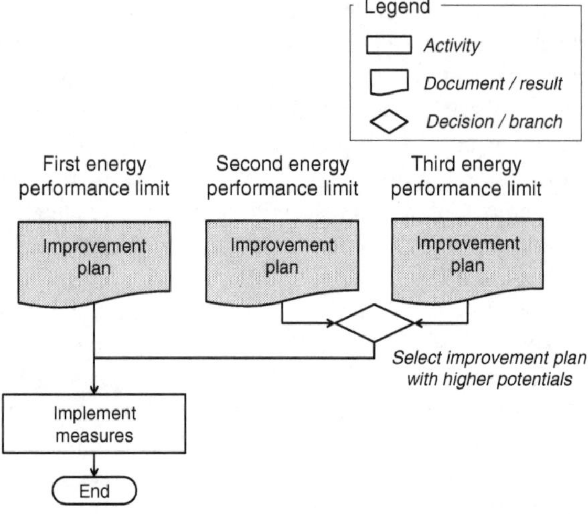

the reduction of the energy demand for machining and the adjustment in the process design. The decision to *implement measures* (*V*) from the available alternatives represents the final activity terminating the improvement process model (see Fig. 4.39). Apart from the implementation of measures to approach the first energy performance threshold, only the improvement plan of the second or third performance limit can be realized in one application cycle. This restriction originates in the necessity to ensure the validity of the original energy consumption function for the third performance reference. This obstructs accordingly any adjustments in the underlying technical capability of the investigated machine tool (see Sect. 4.3.4). As a consequence, the alternative with a higher estimated saving potential represents the preferential choice for implementation together with the means to deactivate the machine tool elements.

Chapter 5
Towards Implementation

This chapter describes the exemplary concept implementation of the energy performance management for machine tools in an industrial case study. First, a concise overview about the initial situation and industry context is given, before the evaluation and improvement in energy efficiency of machine tools is reflected from a user perspective in two applications. These concentrate on the procurement of new machine tools and the improvement of existing systems.

5.1 Initial Situation

The implementation of the developed concept for energy performance management is exemplified based on a collaborative project with the Volkswagen manufacturing site in Kassel. The production plant is one of the key component suppliers of gearboxes within the Volkswagen group. The manufacturing environment is therefore characterized by a multitude and diversity of metalworking processes with over 5,000 currently operating machine tools. The total electrical energy consumption of the plant accumulated in the year 2010 up to 616 GWh, representing a share of 49 % of the total energy demand (Volkswagen 2011b). The manufacturing of more than 3.47 million gearboxes required with 304 GWh nearly 50 % of the plant's electrical energy demand in the same year.

With regard to the intensified economic and ecologic implications of electrical energy usage as well as the continued expansion in the manufacturing capacity towards more than 4 million gearboxes in 2012, the focus of the plant's energy management activities concentrates on the realization of improvements in the energy demand for machine tools. It is challenged to exploit saving potentials among the existing machinery as well as to ensure particularly the installation of new, energy efficient machine tools to satisfy the increasing capacity needs. This background

A. Zein, *Transition Towards Energy Efficient Machine Tools*, Sustainable Production, Life Cycle Engineering and Management, DOI: 10.1007/978-3-642-32247-1_5,

creates the motivation for the prototypical implementation of the concept for energy performance management. It is developed in this context with the objective

- to identify energy efficient machine tools already during the procurement phase,
- to determine energy saving potentials among the existing machine tools in order to guide effort and means towards improvement opportunities.

The energy performance management is specified in both applications following the activities described in the workflow model of Fig. 4.33. The elaborated concepts specialize on certain aspects of the performance management with regard to the requirements of the application and operational opportunities. Due to confidentiality reasons, the data about applied processes and machine tools within the manufacturing system is rendered anonymous.

In order to specify a scope for the energy performance management from the present metalworking process, the relevance of currently applied machining operations at the Volkswagen plant is revised in a technology review. It is furthermore combined with estimates on recent as well as future machine tool demands. The results enable to deduce a machining process of high relevance for the exemplary application of the performance management. The relevance of a technology is estimated in this example based on the available quantities of machine tools and the cumulated power rating. A high availability of systems is positively valuated due to the opportunity to generate economies of scale for the implementation of identified improvement measures. Similarly, the power rating indicates as a second valuating dimension the potential importance of power consumption for the machining technology. The necessary input data is extracted from an inventory list for installed manufacturing processes. Using portfolio analysis, machining centres as well as grinding and turning machines are identified as outliers (see Fig. 5.1). These three processes dominate the other technologies either in availability, cumulated power rating or the combination of both factors. The operation of milling machine tools is also classified as a relevant technology, yet with minor impact than the preceding processes. In contrast, honing and broaching are considered as manufacturing operations with a low relevance for the evaluation and improvement of energy efficiency.

Among the currently dominating technologies, grinding machines are primarily in the focus of recent orders and installations. An analysis of the implemented machining technologies over the last four decades reinforces this impression assigning a substantial share to the material removal process with undefined cutting edges (see Fig. 5.2). Based on the results of the technology review, the concept of energy performance management is subsequently exemplified for grinding machine tools.

The majority of the systems performs outer cylindrical and gear grinding operations emphasising the final processing of hardened steel surfaces. Although the material removal is divergent with regard to specific grinding process, both operations share common characteristics regarding the material properties and general machining task. For that reason, the evaluation of energy efficiency proceeds initially with the assumption that both forms of grinding machine tools are

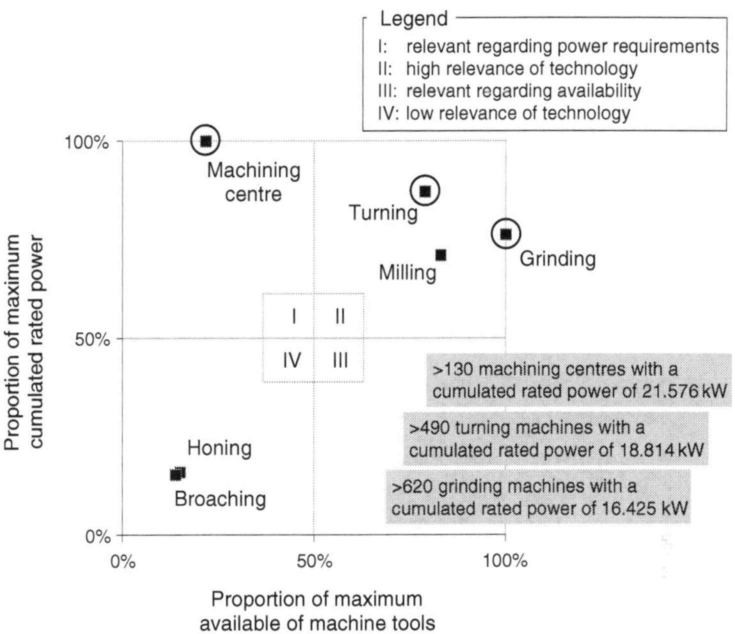

Fig. 5.1 Classification of relevant machining technology

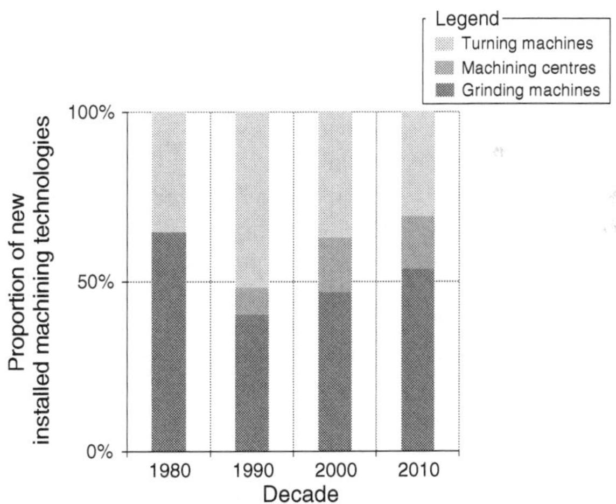

Fig. 5.2 Proportion of installed machining technologies over the last four decades (excerpt of outliers)

comparable from an energy-oriented system perspective. It predefines accordingly the technological boundary for the selection of machine tools in the goal and scope definition of the evaluation process model.

5.2 Procurement of Energy Efficient Machine Tools

Triggered by the necessity to expand and renew the current manufacturing capacity, the number of machine tools is considerably increased with new systems. To seize the benefits of energy-enhanced machine tools, the planning and procurement is challenged to assess the efficiency of energy usage in proposed grinding machine tools. This represents the motivation for the energy performance management in the first example. The application of the concept begins with the formulation of the goal and scope.

5.2.1 Goal and Scope Definition

The objective for the energy performance management is defined as the identification of energy efficient grinding machine tools in the early phases of a procurement process. It relies therefore on the evaluation process model with the process routines to initialize energy limits and quantify energy efficiency gaps (see Sect. 4.7.2). As the focus of the application is on machining technologies, the first and second energy performance indicators are used for the energy performance management.

5.2.2 Initialization of Energy Performance Limits

With the definition of the goal and scope, the activities to initialize the first and second energy performance limits are triggered. As the limit for the energy overhead has a value of zero, the characterization of the energy threshold for the second performance criterion starts with the identification of grinding machine tools to approximate the efficient energy consumption function. In the beginning of the project, 30 entities have initially been determined reflecting commonly used operations and systems for the grinding of hardened steel. The involved grinding machines vary with regard to the year of installation, process parameters and machining tasks. An overview about the basic machine tool characteristics is provided in Table 5.1.

Table 5.1 Determination of grinding machine tools

Characteristic	Specification
Machining tool operation	13 machine tools for gear grinding 17 outer cyclindrical grinding machine
Year of installation	The machine tools have been installed between the years 1998–2009
Process parameters	The material removal rate varies between 2.6 and 50.3 $mm^3\ s^{-1}$
Machining task	The material removal ranges from 120–4,700 mm^3

The identified grinding machines are assessed in the following step regarding the power consumption using the standardized measurement strategy (see Sect. 4.4.3). The energy demand of a machining cycle is determined for each system as an averaged value. Detailed data about the actually removed material volume has also been quantified in experimental measurements. This involved the weighing of parts before and after the machining operation as well as the calculation with the specific material density of the applied steel. Together with the material removal rate, all necessary parameters to characterize the machining activities are collected. The specific energy demand and the material removal rate for the 30 machine tools are illustrated in Table 5.2 representing the input data for the identification of efficient operations in the following step.

Based on these aggregated parameters, each machining process is reviewed in order to identify the efficient activities (see Fig. 5.3). These dominate the energy

Table 5.2 Aggregation of measured values for the investigated machining activities

Machine tool	MRR ($mm^3\ s^{-1}$)	SEC ($Wh\ mm^{-3}$)
1	2.6	0.7
2	2.8	0.56
3	3.0	0.8
4	3.3	0.47
5	3.7	0.37
6	8.7	0.43
7	10.6	0.11
8	11.0	0.24
9	11.3	0.16
10	12.3	0.20
11	13.9	0.11
12	14.4	0.09
13	14.7	0.13
14	19.7	0.18
15	20.7	0.12
16	22.4	0.09
17	23.8	0.13
18	24.9	0.09
19	26.2	0.07
20	28.0	0.06
21	28.9	0.10
22	31.5	0.10
23	32.1	0.11
24	33.2	0.08
25	37.0	0.04
26	38.3	0.06
27	41.7	0.03
28	44.9	0.09
29	45.9	0.07
30	50.3	0.04

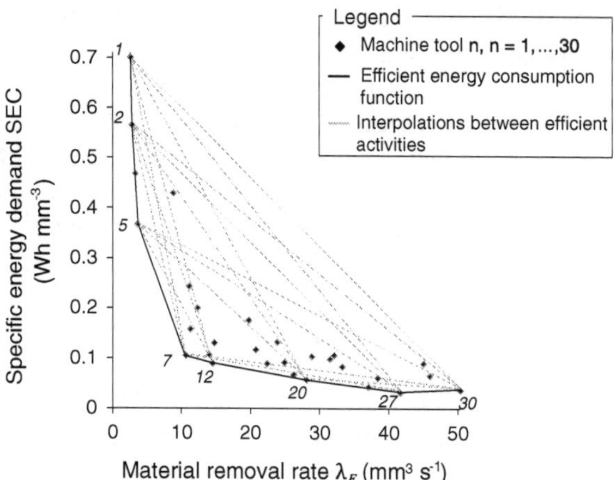

Fig. 5.3 Resulting efficient energy consumption function for the grinding machine tools

Table 5.3 Linear segments enveloping the efficient consumption function

Segments	Linear equation	Application range
1–2	$a_{E1,2} = -0.7205\lambda_E + 2.5998$	$2.64 \leq \lambda_E \leq 2.82$
2–5	$a_{E2,5} = -0.2238\lambda_E + 1.197$	$2.82 \leq \lambda_E \leq 3.71$
5–7	$a_{E5,7} = -0.0383\lambda_E + 0.5093$	$3.71 \leq \lambda_E \leq 10.55$
7–12	$a_{E7,12} = -0.0038\lambda_E + 0.1453$	$10.55 \leq \lambda_E \leq 14.43$
12–20	$a_{E12,20} = -0.0024\lambda_E + 0.1254$	$14.43 \leq \lambda_E \leq 28.03$
20–27	$a_{E20,27} = -0.0018\lambda_E + 0.1068$	$28.03 \leq \lambda_E \leq 41.65$
27–30	$a_{E27,30} = 0.0006\lambda_E + 0.0083$	$41.65 \leq \lambda_E \leq 50.29$

requirements of alternative systems operating at the same material removal rate as well as the linear interpolations between these activities. As a result of the analysis, eight processes are determined, which comply with the requirements of the dominance criterion. The identified entities are therefore assigned as efficient. The linear interpolations between these activities are then determined to approximate a frontier of dominating enveloping functions. These segments characterize the efficient energy consumption function for the revised machining technology.

The enclosing segments of the consumption function are mathematically calculated in the final step. The linear equation for each function is determined based on the two involved efficient activities (see also Appendix C). The resulting functional relations are concluded in Table 5.3. Depending on the applied material removal rate, these equations indicate a minimum attainable specific energy demand for the considered grinding technology. Together with the indicator regarding the energy overhead, the derived performance limits finalize the activities to initialize the energy performance limits for the grinding machine tools.

In order to revise the assumption on the comparability of outer cylindrical and gear grinding operations, the consumption functions for the individual machine tool processes have additionally been differentiated. The enveloping functions are

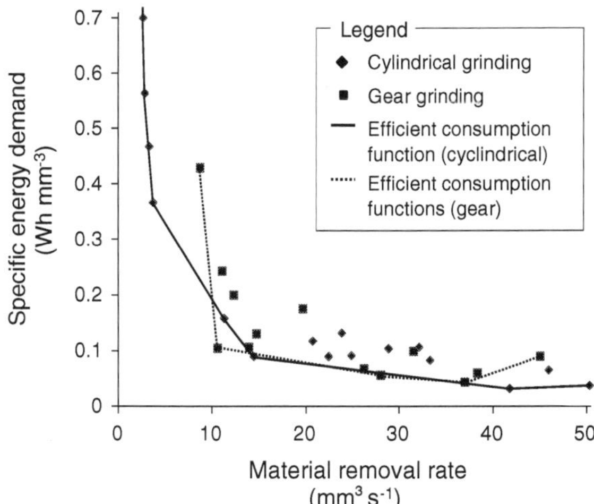

Fig. 5.4 Comparison of efficient consumption functions for gear and cylindrical grinding

visualized in Fig. 5.4 revealing a high conformity between both frontiers over a wide processing range. For that reason, the evaluation of energy efficiency proceeds initially with the assumption that both forms of grinding machine tools are comparable for the prevalent machining task from an energy-oriented perspective.

5.2.3 Quantification of Energy Efficiency Gap

Using the ascertained performance limits for grinding machine tools, the assessment of the energy efficiency gap for the procurement of machine tools is elaborated in accordance with the workflow described in Fig. 4.35. A challenge for the evaluation of the actual and ideal energy performance in the early phases of procurement is especially that data about the energy requirements of a proposed machine tool is not directly quantifiable by the user. It demands therefore to establish an interaction with machine tool manufacturers to ensure the determination and provision of energy-relevant data. As a consequence, an energy-oriented procurement process is designed. It divides the activities to quantify the energy efficiency gap for machine tools between the user and manufacturer (see Fig. 5.5).

Following the derivation of energy limits, the user submits for a planned machine tool procurement the energy limits as part of the technical specifications requesting feedback information about the projected actual energy performance. The observation of the performance is then conducted by the manufacturer for targeted process conditions using the standardized measurement strategy introduced in Sect. 4.4.3. The requested data enables the user to compare the current performance with the associated energy limits and to determine the divergence between the indicator values.

The evaluation of energy efficiency for three actually proposed machine tools is visualized in Fig. 5.6. The actual performance is compared against the first and

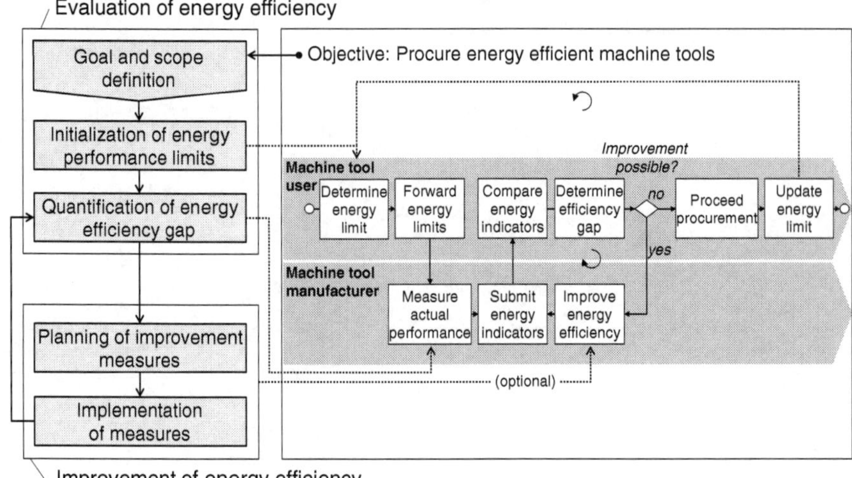

Fig. 5.5 Integration of energy performance management in the procurement process for machine tools

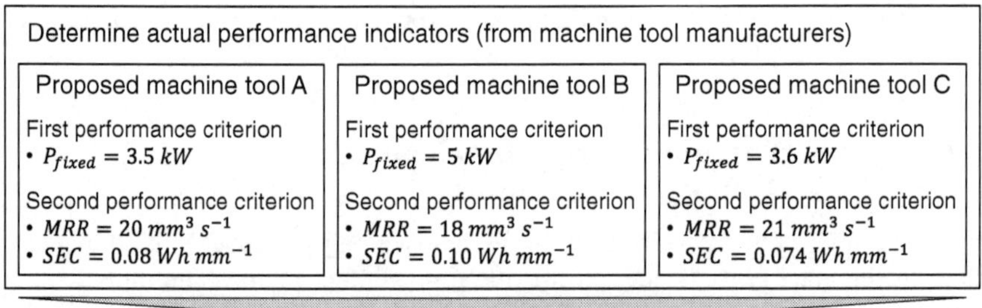

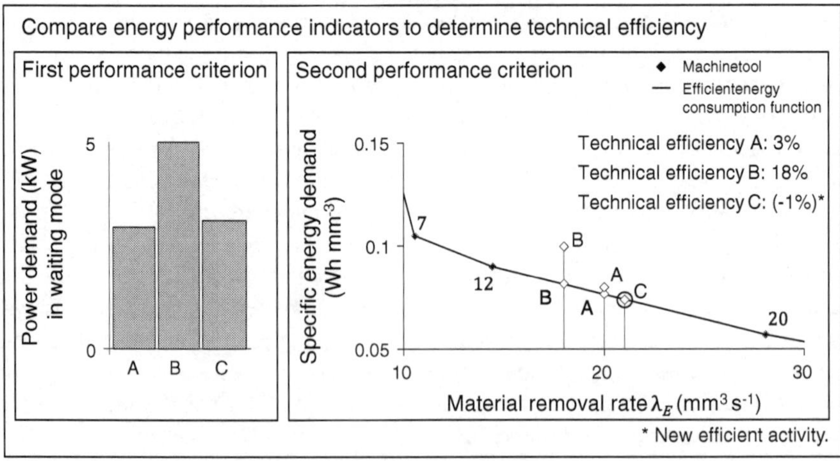

Fig. 5.6 Quantification of energy efficiency gaps for proposed machine tools

second energy limit to quantify the efficiency gap. The results show that machine tool A has the lowest power demand in waiting mode compared to the other systems. Considering the second performance criterion, the divergence between the specific energy demand of the actual unit and the corresponding reference limit is calculated (in this case equation $a_{E_{12,20}}$). The results reveal that machine tool C is superior in the second energy performance criterion to the alternatives A and B. It is even more declared as efficient exceeding the current reference level.

Based on the resulting energy efficiency gaps, the machine tools A and C are identified as preferential solutions. Yet, none of the machine tools is dominating in both criteria. This necessitates assessing the priority of the first and second energy limit for instance. A possible prioritization is for instance to revise the impact of both performance indicators on the absolute energy demand for an anticipated operational behaviour. Figure 5.7 visualizes the machine tool with the lower total energy demand in dependence of the total removed material volume and waiting hours per day. The results illustrate that the lower specific energy requirements of machine tool C exceed the enhanced fixed power demand of machine tool A in cases of minor waiting times and high material removal, thus indicating the preferential choice for the high-scaled automotive application.

To enhance the determined technical efficiency of the proposed machine tools, the review in the early phases of procurement opens up the opportunity to implement a control loop triggering communication between the user and manufacturer about the origins of inefficiency and potentials. Using optionally the methods and tools of the improvement modules by the manufacturer, a re-examination of the energy efficiency gap is conducted for an updated proposal. This applies in particular for all three revised machine tools regarding the zero energy overhead of the first energy performance criterion.

The machine tools with its determined technical efficiencies are finally documented as part of a procurement proposal enabling to compare the energy efficiency

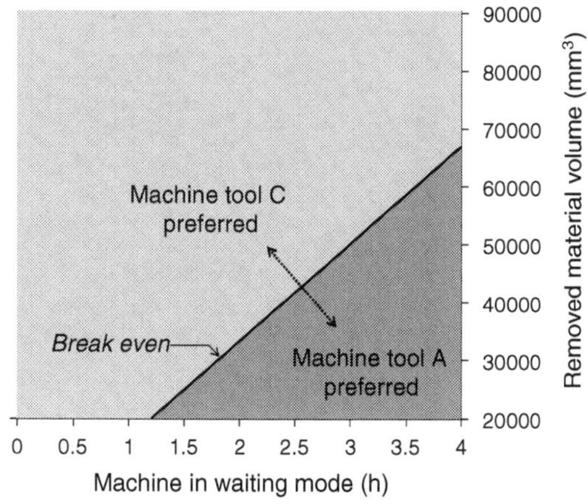

Fig. 5.7 Identification of the preferred machine tool in consideration of the first and second energy performance limit

together with technical and economic factors for the final decision-making. Prior to terminating the energy-oriented procurement process, the energy consumption function is updated integrating the identified efficient system C. This ensures to maintain a contemporary reference dataset and revise technical progress of the technology (see Fig. 4.17).

With regard to the formulated need to identify energy efficient machine tools, the introduced approach describes a quick and lean solution to analyse and compare the energy requirements of planned systems in the early phases of procurement. Revising the fixed power consumption and specific energy demand, it provides transparency about the essential energy requirements in operating and non-operating mode. The quantification of the energy efficiency gaps enables furthermore an immediate comparison among proposed machine tools. This eases the consideration of the energy demand as one element of the multiobjective decision problem within machine procurement. Especially by forwarding the energy limits and implementing the succeeding feedback loop, important opportunities are provided to adjust the energy requirements already in the planning and design phase towards efficient machine tool systems. The integration of the energy performance management into this planning process indicates yet additional benefits with regard to the application of the energy consumption function.

In contrast to the final target of a zero energy overhead, the efficient energy consumption function has an adaptive characteristic enabling to observe adjustments of the frontier. Within the procurement process, it is primarily used as an object of comparison serving the function to contrast the actual energy performance of a planned machine tool towards a reference. Despite the passive use of the consumption function, it can conversely be applied as an active element in the functional specification and encourage technical progress in the energy requirements of machine tools at the beginning of the procurement. Figure 5.8 visualizes three

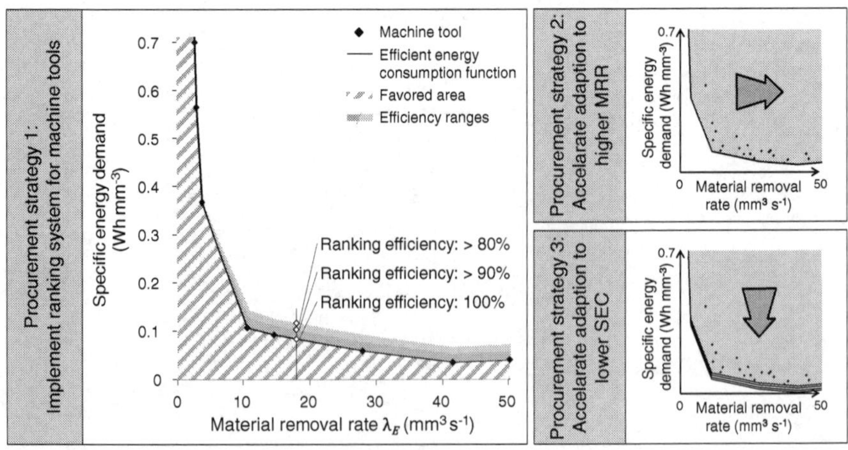

Fig. 5.8 Formulation of procurement strategies for machine tools

future procurement strategies differentiating the ranking of machine tools as well as the adaption to higher material removal rates and lower specific energy demands.

Adapting energy labelling schemes, the efficient consumption function provides the opportunity to define a favoured area and a tolerance range of inefficiency for future machine tool procurements. These ranges enable to categorize machine tools according to their technical efficiency using efficiency classes as introduced for electrical motors (e.g. IE1 to 4) (de Almeida et al. 2008; Herrmann et al. 2007). Creating transparency and comparability among different systems, it facilitates to gain instant awareness about the energy requirements of machine tools and serves as a stimulus to increase energy efficiency.

With regard to general contour of the efficient energy consumption function, two additional strategies can be formulated stating objectives for the procurement of future machine tool generations. These include on the one side to strive for higher material removal rates in order to benefit from the reduced specific energy demands. On the other side, a relative improvement in relation to the status quo can be promoted encouraging machine tool manufacturers to enhance continuously the energy requirements with each new machine tool development (Schischke et al. 2012a).

5.3 Evaluation and Improvement of Energy Efficiency for Existing Machine Tools

The Volkswagen manufacturing site in Kassel operates currently more than 620 grinding machine tools. Apart from the procurement of new, efficient machine tools, it is indispensable to revise improvement opportunities among the existing systems. This represents the motivation for the energy performance management in the second example. The application of the concept begins with the formulation of a goal and scope.

5.3.1 Goal and Scope Definition

The aim for the energy performance management is defined as the identification of saving potentials among existing machine tool systems and the derivation of means for improvement. It relies therefore on the evaluation and improvement process models (see Sect. 4.7.1). Maintaining a focus on machining technologies, the first and second energy performance indicators are used again for the energy performance management.

The scope and technological system boundary build upon the results of the preceding application. This includes the energy consumption function for the evaluation of energy efficiency as well the actual performance values for the 30 grinding machine tools. As a consequence, the quantification of energy efficiency starts directly with the activity to determine the technical efficiency of the revised systems.

5.3.2 Quantification of Energy Efficiency Gap

In the previous power measurement for the 30 grinding machine tools, the power demands in waiting mode as well as the specific energy consumption (SEC) at the given material removal rate have been determined. The technical efficiency for the first energy performance limit represents the actual power demand in waiting mode. Alternatively, the efficiency gap for the second energy performance limit constitutes the divergence in the SEC for the actual activity and corresponding efficient reference on the consumption function (see Fig. 5.9). The resulting quantities for each grinding machine tool are illustrated in Table 5.4. The calculation of the technical efficiency is also provided in detail in Appendix D.

The analysis of the results shows that the first energy limit is not achieved by any of the observed machine tools (see Fig. 5.10). The measured power demands in waiting mode vary between 2.8 and 10.6 kW with an average of 4.96 kW. With regard to the second energy performance limit, the technical efficiency indicates a relative improvement potential from 2 up to 61 %. The average efficiency gap between the ideal and actual energy performance for the revised grinding machine tools has a value of 37.8 %.

The measurement results reveal in addition a substantial inefficiency for machine tool 29 evincing the highest improvement potential among all studied grinding machine tools. It has the highest power demand in waiting mode among all observed units contributing substantially to the SEC. The fixed power demand provides accordingly an indication about the origins of the corresponding technical efficiency in the second energy performance limit. Conversely, the machine tools maintaining a low power demand in waiting mode seem to be also featured

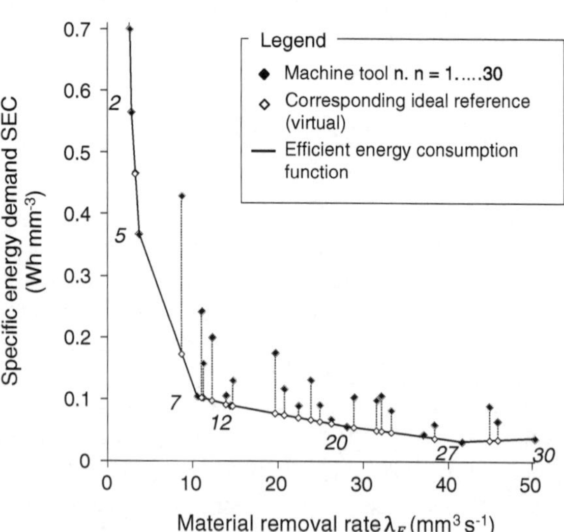

Fig. 5.9 Determination of technical efficiency for the second energy performance limit

Table 5.4 Technical efficiencies for the examined grinding machine tools

	First energy limit	Second energy limit
Machine tool	Technical efficiency (kW)	Technical efficiency (%)
1	4.0	0
2	4.4	0
3	4.8	35
4	3.1	2
5	4.1	0
6	5.1	59
7	3.2	0
8	7.8	58
9	4.7	35
10	6.5	51
11	4.2	13
12	3.2	0
13	5.9	31
14	5.6	56
15	4.0	36
16	5.0	21
17	4.9	48
18	5.1	28
19	3.0	9
20	2.8	0
21	5.6	48
22	5.7	50
23	5.1	54
24	5.8	44
25	3.7	10
26	6.5	38
27	3.6	0
28	10.6	61
29	4.7	45
30	4.3	0

with a low technical efficiency for the specific energy demand. With the generation of these datasets for evaluation, the process routine to determine the efficiency gap is terminated triggering the improvement of energy efficiency for selected machine tools.

Prior to the initiation of the process routines for improvement, the derived data provides additionally a distinct insight into development trends in the technical efficiency. The actual performance values of three units from the available collection of machine tools are illustrated in a comparative assessment of the power demand in waiting mode and the SEC (see Fig. 5.11). The units 7, 8 and 20 vary with regard to the conceptual machine tool design, but carry out the same operations providing a distinct example for the comparison of machine tools with an indistinguishable technological scope.

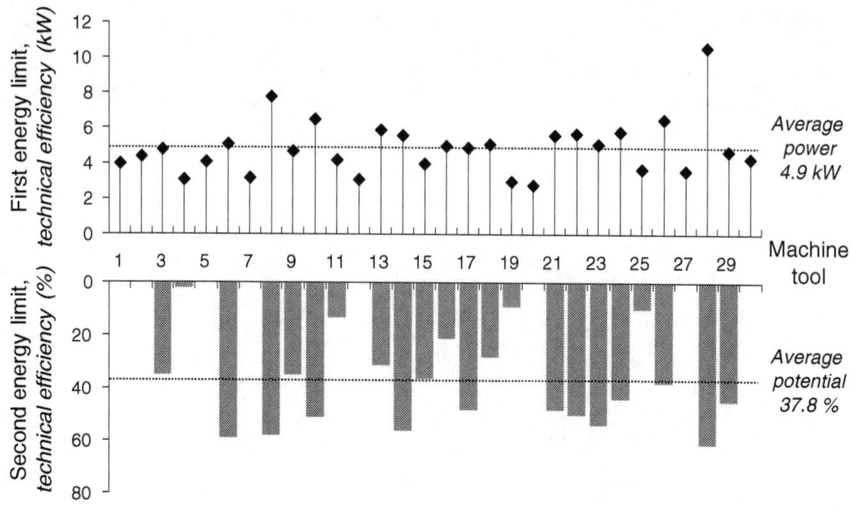

Fig. 5.10 Analysis of technical efficiencies for the grinding machine tools

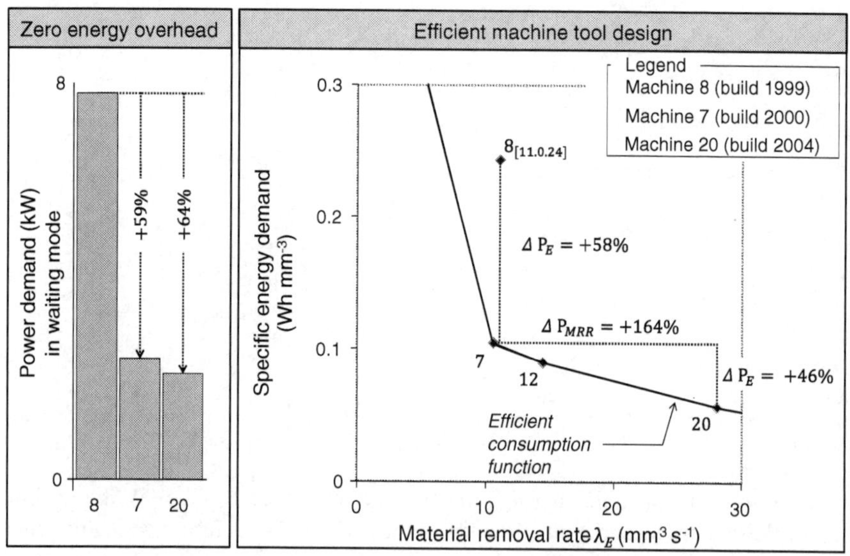

Fig. 5.11 Relative technical efficiencies of three machine tools with the same machining task

A comparison of the relative technical efficiency between the three machine tools shows that the actual performance values in both energy performance indicators differ substantially. Considering the first performance criterion on the energy overhead, the power demand in waiting mode is enhanced between machine tools 8 and 7 by 59 %, in relation to the machine tool 20 even up to 64 %. Following the alteration of the second energy performance indicators according to the year of installation, a deviation of 58 % in the specific energy demand (SEC) is determined between machine tools 8 and 7. Towards machine tool 20, an additional improvement in the SEC by 46 % is observed relative to machine 7. While the machine tools 7 and 8 operate with an approximate material removal rate, it is increased by 164 % for machine tool 20. The origins for the enhanced SEC can accordingly be presumed in the acceleration of the material removal. The example underlines again the impact of the fixed power consumption causing a high SEC. It clarifies even more the dynamic potential of technical developments, which enabled to reduce the energy requirements to conduct the material volume between machine tools 8–20 by 77 % within 5 years. By reducing the fixed power demand and increasing the material removal rate, the comparison of the three machine tools enables accordingly to conclude two general means in order to enhance the actual energy performance towards the ideal reference.

With the determination of the technical efficiencies for 30 machine tools, an initial overview is provided about the prevalent improvement potential regarding the first and the second energy performance limit. The range of available values for both performance indicators underlines the diversity of the concepts and possible design solutions among the considered machine tools. In accordance with the formulated objective for the energy performance management, the quantification of the energy efficiency gaps enables instantly to identify and prioritize machine tools providing guidance for the initiation of means for improvement.

5.3.3 Planning of Improvement Measures

Based on the evaluation of energy efficiency, the first and foremost improvement measure is initially to use the efficient machine tool. The comparison of machine tool 8 and 20 illustrates the attainable saving potential, thus promoting to utilize the efficient machine tools to the full capacity and use inefficient systems as buffer capacity. For the particular application, the allocation of all machining tasks from machine tool 8–20 accumulates to a total energy saving of more than 22,000 kWh/a.

Further opportunities and potential means to enhance the technical efficiency of existing grinding machines are revised subsequently using the methods and tools of the concept modules energy breakdown analysis and improvement planning (see Sect. 4.7.2).

The concurrent power and machine data analysis is applied for a selection of machine tools monitoring the power requirements and operation of involved electrical components. A power log chart for machine tool 20 is provided in Fig. 5.12. The observed activation of the involved machine tool components corresponds generally with the revised operational characteristic of the elements in other systems. The analysis of the power log chart indicates that the tool spindle (b-axis) is generally activated throughout the entire monitoring period including an enforced waiting period of several minutes. Conversely, the other axes remain inactive as long as the machine tool is not involved in a machining cycle.

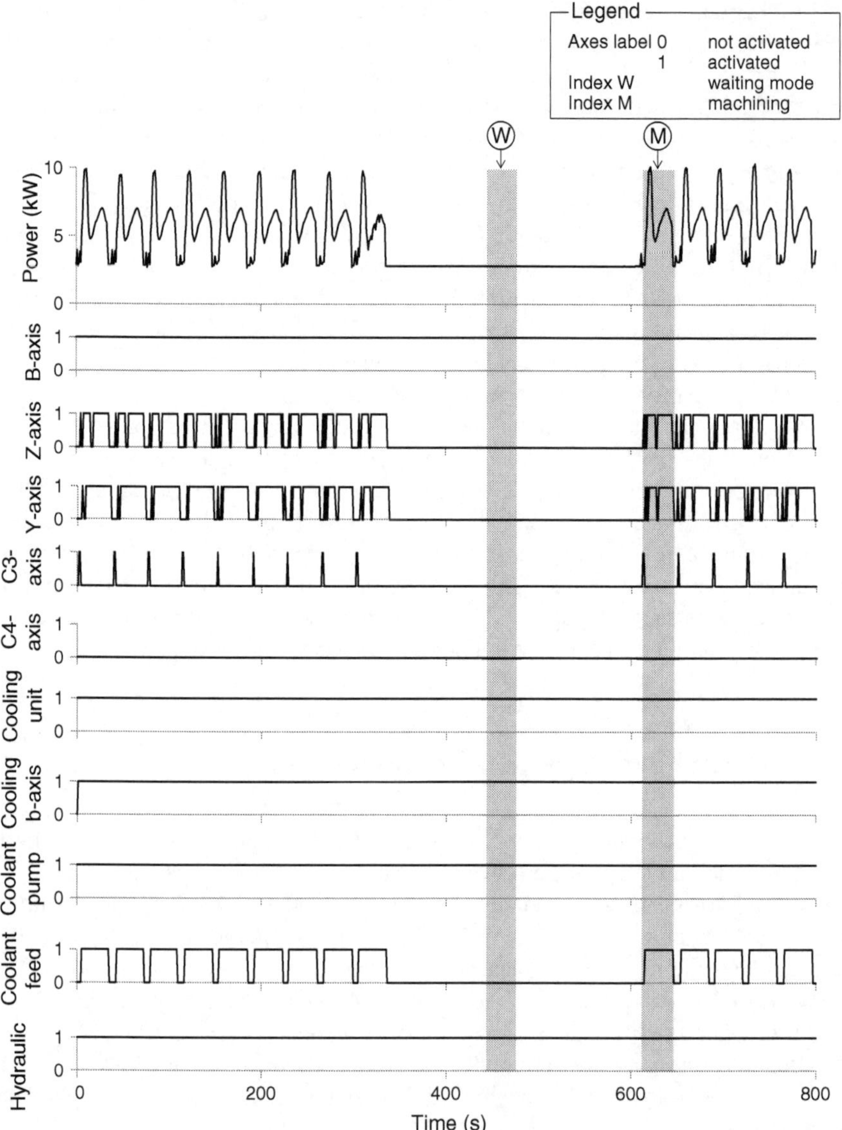

Fig. 5.12 Power log chart for the concurrent power and machine data analysis of the grinding machine tool 20

Combining the activated components in waiting mode and during machining with the associated input power rating, a ranking is deduced for machine tool 20 (see Table 5.5). It points out the high relevance of the spindle and cooling unit followed by the z- and y-axis during machining and the hydraulic in waiting mode. As the results of the power log chart do not enable to approximate distinctively the power consumption in this case, selective power measurements have been conducted on the

Table 5.5 Ranking of relevant components for the grinding machine tool

Component (rated input power)	Normalized rated power (%)	Active in machining (%)	Prioritization "machining"	Active in waiting mode (%)	Prioritization "waiting"
Spindle B-axis (13.75 kW)	100	100	A (100 %)	100	A (100 %)
Z-axis (1.4 kW)	10	90	C (9.0 %)	0	–
Y-axis (1.4 kW)	10	90	C (9.0 %)	0	–
C3-axis (1.75 kW)	13	10	E (1.2 %)	0	–
Dresser C4-axis (1.4 kW)	10	0	–	0	–
Cooling unit (1.75 kW)	13	100	B (13 %)	100	B (13 %)
Hydraulic (0.3 kW)	2.0	100	D (2.0 %)	100	C (2.0 %)

component level. These confirm on average the estimated prioritization of machine tool elements defining a structured order for the review of improvement measures.

The results of the individual power measurements for components enable once more to illustrate changes in the component design using the three revised grinding machine tools 8, 7 and 20. While the hydraulic unit represents for instance the relevant component for machine tool 8, the drive system (including the spindle) is considered with a high priority for machine tools 7 and 20. The average power requirements of both components are compared in Fig. 5.13 for all three machine tools. The results point out a substantial reduction of the power consumption for

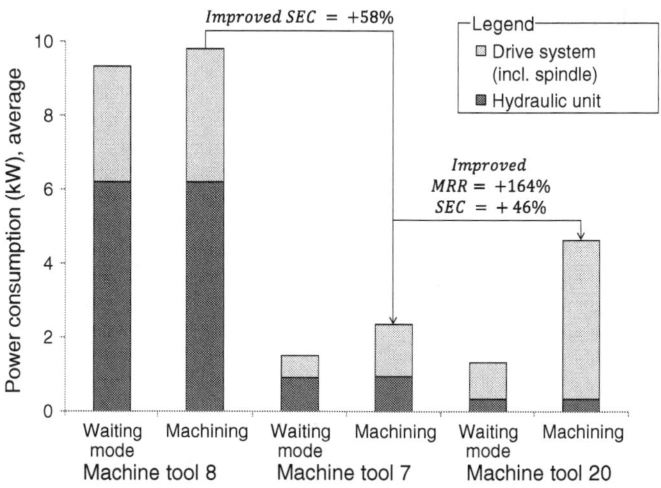

Fig. 5.13 Contrasting the average power consumption of the drive system and hydraulic unit for the machine tools 8, 7 and 20

the hydraulic diminishing from 6.2 kW for machine tool 8–0.9 kW for system 7 to finally 0.3 kW in system 20. Considering the deviation of the power consumption for the different drive systems, especially an increased demand due to the higher material removal rate can be identified. The example of machine tool 20 underlines consequently the complexity to deduce improvement measures. It is generally not advantageous to restrict the power demand of a machine tool element, if this limits an improvement of the whole system (Stasinopoulos et al. 2009). Integrating for instance a spindle with a high technical performance and correspondingly high power consumption, machine tool 20 is enabled to realize a higher material removal rate thus compensating the additional power demand by a reduced process time.

On the basis of the power log chart and the comparison of the power demands for the three grinding machine tools, starting points for the improvement of the machine tools can be deduced. These encompass the reduction of the power demand in waiting mode as well as the enhancement of the SEC through process adjustments and the reduction of energy demands. The planning of improvement is conducted in accordance with the process routine visualized in Fig. 4.37. Based on the results of the energy breakdown analysis, measures have been assigned to meet the derived opportunities. In order to realize improvement in the actual performance value of the first and second energy criterion, a set of selected measures is initially compiled, which are applicable to the majority of the considered grinding machine tools.

With an average power consumption of 4.9 kW in waiting mode, the first measure on the list is assigned to the deactivation of components in non-operating mode. Revising the power log chart of the machine tool 20, the measure focuses initially on the deactivation of the spindles in each machine tool. The controlled shut-down of further components necessitates the cooperation with the machine tool manufacturer to define a turn-off strategy under the premises to realize greatest possible energy savings without compromising the availability of the machine tool for machining. The results of an experimental analysis of waiting intervals in

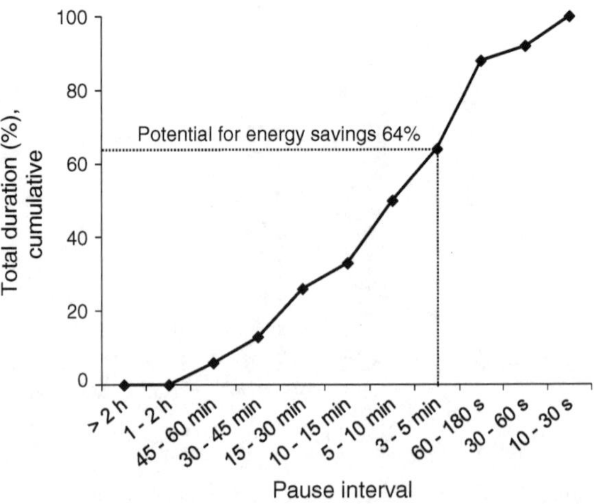

Fig. 5.14 Distribution of idle times in automotive production cells (PROFIBUS Nutzerorganisation e. V., 2011)

automotive production cells provide guidance for the design of the deactivation routine (see Fig. 5.14). These designate a beneficial temporal threshold to deactivate components for waiting intervals exceeding 5 min. This enables for instance in the revised case to save 64 % of the energy demand in non-operating times. As an indicative value for the assessment of profitability, the saving potential for a grinding machine tool can be estimated with 5,100 kWh/a assuming the fixed power demand of 4.9 kW and a machine availability of 80 % (referenced to 220 working days and three operating shifts).

Based on the specification routine to deduce improvement measures for the second energy performance criterion, two options are proposed for implementation striving to reduce the time demand of the machine tool system. The first one addresses the acceleration of the material removal rate (MRR). The realization of this measure has been tested for machine tool 20 increasing the MRR from 11–14 $mm^3 s^{-1}$. The resulting saving potential of 32 % in the SEC accumulates up to more than 17,000 kWh/a under the given machining conditions. Prior to adjustments in the material removal rate, it is essential to test the general feasibility of the planned process condition in order to avoid negative impacts on the material surface quality as well as the machine and process capability.

An alternative means to reduce the time demand for machining relates to the removed material volume. Transferring the removal of material to less energy demanding operations at the beginning of the process chain (e.g. turning and milling), a reduction of material removal by 31–34 % for the grinding processes has been demonstrated in an experiment (see Fig. 5.15). Under the given conditions, the saved energy demand obtained a value of 5,200 kWh/a over the entire process chain.

In addition to the approaches for the system level, potentials on the component level have been revised according to the feasibility to adjust, dimension or substitute machine tool elements ranging from the drive system and hydraulic unit towards low watt valves. Based on the prioritization of the energy

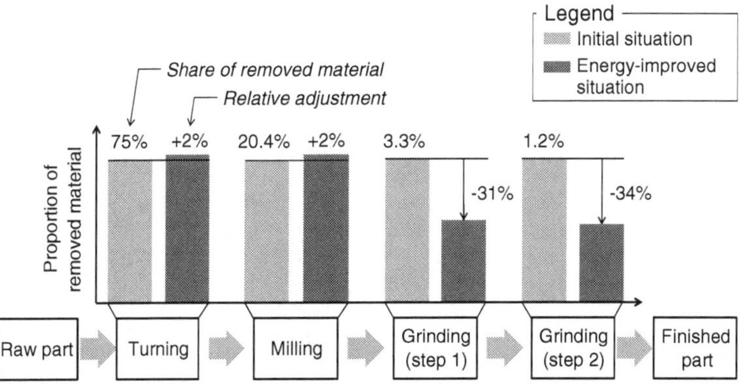

Fig. 5.15 Reduction of removed material volume for grinding

breakdown analysis, this involved initially the use of the Siemens Sizer to evaluate saving potentials by analysing the design of spindles and drives. Among the reviewed machine tools, the identified energy savings amounted for instance up to 600 kWh/a for a spindle drive, thus indicating only a minor potential to initiate the substitution of an entire spindle drive system. All in all, the improvement of components has generally been connected with a minor energy reduction potential among the revised machine tools. This is probably due to the necessity to adapt an enhanced component to the existing machine tool system limiting the design opportunities to realize the full potential. Based on the results of the improvement planning, an exchange of information with the machine tool manufacturers is initiated to continue the identification of means for improvement among existing systems.

The application of the energy performance management to determine energy saving potentials among the existing systems revealed distinct means to enhance the actual performance. With the quantification of the technical efficiency among existing machine tools, an essential opportunity is provided to allocate the machining tasks on efficient machine tools. Concluding the results of the improvement planning, the effort to enhance the actual performance of existing machine tools is guided towards the reduction of the power demand in waiting mode as well as the adjustment in the operation of the machine tool system. The realization of the proposed measures depends especially on the estimated individual saving potentials and the implementation costs. In the end, the final decision to realize a measure is built upon a case-by-case analysis revising the current profitability to initiate the implementation for an existing system. The complexity to revise means for improvement among existing machine tools emphasizes again strongly to use the opportunities provided in the procurement of machine tools in order to avoid subsequent adjustments.

Chapter 6
Summary and Outlook

This final chapter provides a summary on the developed concept of energy performance management for machine tools and concludes the resulting findings of the present research. It includes an examination on the innovative character of the introduced approach and delineates particular restrictions to be considered for the application of the concept. The chapter concludes finally potential consecutive research extending the scope of this work.

6.1 Summary

The threat of environmental change and the impeding scarcity of non-renewable energy resources enforce the decoupling of economic growth and energy-related emissions in industrial economies. Energy efficiency represents the most cost-effective and immediate strategy of a sustainable development enhancing the required energy demand of transformation processes to generate a designated output. Due to substantial energy-related environmental and economic implications, a strong emphasis is currently put on the electrical energy requirements of operating machine tools for metalworking processes. It is enforced by the initiation of regulatory frameworks as the Ecodesign directive as well as an increasing awareness and consideration of energy demands as part of the life cycle costs by machine tool users. The improvement of energy efficiency for machine tools is however confronted with diverse barriers, which sustain an energy efficiency gap of unexploited potential. The deficiencies lie in the lack of information about the actual energy requirements of machine tools, a minimum energy reference to quantify improvement potential and possible actions to improve the energy efficiency.

Bridging the energy efficiency gap represents for that reason the motivation for this research. The primary objective has been the development of an energy performance management concept, which provides methods and tools to quantify and

A. Zein, *Transition Towards Energy Efficient Machine Tools*, Sustainable Production, Life Cycle Engineering and Management, DOI: 10.1007/978-3-642-32247-1_6,

evaluate the energy efficiency of machine tools as a basis to initiate improvement towards an ideal energy level.

To achieve the objective, the fundamental context for improving the energy efficiency of machine tools has been introduced in Chap. 2. This involved revising the origins of energy usage and measures for improvement as well as introducing performance management as a management system to uncover improvement potential and pursue progress. Although performance management provides a consistent framework of methods and tools, a lack of distinct approaches to determine the divergence between an actual and minimum energy demand of machine tools was found.

A review on the state of research has been conducted in Chap. 3 in order to identify approaches, which provide a contribution to the defined objective. Striving to bridge the energy efficiency gap of machine tools, this included especially methods and tools to quantify, evaluate and improve the energy demand. As a result of the comparative assessment, an ensuing need was deduced for a comprehensive, conceptual approach to evaluate the energy efficiency of machine tools and guide the improvement towards an unequivocal, minimum energy demand.

In line with the identified need for research, the developed concept for energy performance management of machine tools has been elaborated in Chap. 4. It is composed of four concept modules, which are interconnected in a workflow model to evaluate and improve the energy efficiency of machine tools:

- Within the concept module to *determine energy performance limits*, methods and tools are provided to deduce three technically achievable energy values as a reference to quantify the energy efficiency gap. Adopting a systems-based perception of an ideal energy demand for transformation processes, these limits obtain both a machine tool and process perspective deriving minimum threshold values for the fixed power consumption and the process energy demand.
- The *measurement of the actual energy performance* defines a standardized measurement strategy to ensure a consistent and effective observation of the actual performance for an investigated machine tool. It includes especially the specification of unified routines for the experimental power measurement. The energy limits and the corresponding actual energy performance values represent the essential information to quantify the energy efficiency gap.
- The concept module on *energy breakdown analysis* introduces a universal methodology supporting the conversion of a quantified energy efficiency gap for complex machine tool systems into improvement means. It establishes a concurrent logging of the temporal operation and power demand of integrated machine tool elements in order to provide structured guidance on the identification and exploration of improvement opportunities.
- The *improvement planning* targets the enhancement of actual energy performance providing methods and routines to allocate improvement measures to the identified potentials and formulate improvement policies. This includes the assignment of measures to the energy performance limits structuring their review and specification.

- The application of the four preceding concept modules is structured in *workflow models*, which delineate the processes to evaluate and improve the energy efficiency for machine tools. The workflow models are equipped with feedback mechanisms in order to implement an iterative operation of the evaluation and improvement establishing a continuous improvement process.

The applicability and capability of the energy performance management of machine tools in an industrial case study has been demonstrated in Chap. 5. The developed concept was implemented in an automotive manufacturing environment with the aim to identify energy efficient machine tools already in the early phase of procurement and to determine energy saving potentials among existing machine tools. As a result of the prototypical concept realization, energy performance limits have been derived for grinding machine tools. These were initially used to set up a procurement process easing the energy-oriented assessment of proposed grinding machine tools. Regarding the review of improvement potentials among the existing machine tools, the energy performance management revealed substantial saving potential regarding the fixed power demand and process energy demands guiding the effort to initiate improvement and bridge the efficiency gap.

This work concludes in this chapter with a critical appraisal of the innovative contribution of the developed concept to the identified need for research. The introduced concept for energy performance management provides for the first time a comprehensive approach providing methods and tools to overcome the barriers, which obstruct the transition towards energy efficient machine tools. Reflecting the findings and particular restrictions to be considered for the application of the approach, an outlook on consecutive fields of research completes this work. It guides further research specifically towards the characterization of technological system boundaries for machine tools from an energy-oriented perspective delineating the scope and criteria for the comparison of machine tool processes.

6.2 Critical Review

The introduced concept for energy performance management represents a response to the identified ensuing need for action to bridge the energy efficiency gap for machine tools, which has been deduced based on the comparison of the state of research in Sect. 3.3. For the first time, a comprehensive approach is described, which enables to overcome the barriers obstructing the transition towards energy efficient machine tools.

In contrast to existing approaches, the innovative character of the concept lies in the provision of a novel methodology to describe technically minimum achievable threshold values for the electrical energy demand of machine tools. These provide guidance on the design and operation of energy efficient systems and represent a reference to evaluate the actual suitability and performance capacity of machine tool concepts. In this way, it establishes for the first time the opportunity to quantify the energy efficiency gap as a prerequisite to initiate the exploitation of an absolute attainable improvement potential for individual machine tools.

Beyond the evaluation of energy efficiency, the introduced performance management concept provides new methods and tools to analyse the origins of the actual energy requirements and operationalize identified improvement potentials into distinct measures and improvement plans. Contrary to prevailing approaches, it obtains initially a systems perspective providing a structured guidance on the review of potentials and the assignment of measures among the elements of complex machine tools.

Despite the single focus on the quantification, evaluation or improvement of energy demands, the developed energy performance management encourages for first time the combination of all aspects in a structured, systematic management process. In mind of the barriers obstructing the improvement of machine tools, the application of the concept is formalized in generic workflow models sustaining lean and quick process routines to gain evaluative information and initiate improvement effort for a machine tool manufacturer and user.

The outlined concept for energy performance management presents for the first time a consistent, comprehensive approach guiding the evaluation and improvement of the energy usage in machine tools towards a technically achievable energy minimum. Throughout the application of the introduced concept, the following critical aspects have to be considered to ensure an effective transition towards energy efficient machine tools:

- *Determination of efficient energy consumption function*. The formulation of the second energy performance limit is based on an empirical approximation of existing machine tools. The enveloping frontier of the efficient energy consumption function is therefore highly dependent on the selection of machine tools, which are considered in the initial assessment. As a consequence, the resulting consumption functions from alternative sources can differ in the estimation of an efficient reference frontier for the same technology. However, the adaptive characteristic of the efficient energy consumption function enables in the current development already to harmonize these deviations among alternative frontiers by encouraging the continuous expansion and integration of new efficient machine tools.
- *Scope of technology*. The efficient energy consumption function relies on the comparison of machine tools with a comparable technological scope. The definition of a boundary for a technology is initially challenged to outline unifying characteristics, which enable to assign a manufacturing process based on a fulfilment of the given criteria. The characterization of the technological scope relies in the current approach on empirical observations and experience. For that reason, the expansion of the efficient energy consumption function to other manufacturing processes demands always to revise and define technological boundaries keeping in mind to retain the greatest possible scope necessary for a distinct technology.
- *Realization of the energy breakdown analysis*. The energy breakdown analysis is used an interim step within the developed approach to support the conversion of the technical efficiency for the machine tool system to the integrated electrical elements revising their operation and power demands. The results of the

concurrent power and machine data analysis are generated using a prototypical application, which imposes to identify manually the individual component signals for each machine tool and establish a data connection to the control. However, the potential to gain awareness about the operation and potential power requirements has been demonstrated encouraging the implementation especially for complex machine tool systems with several moving axes.
- *Assignment of improvement measures*. The specification of improvement measures provides primarily structured guidance for the review and identification of distinct means to enhance the actual performance of the second energy criterion. The defined routine and generic measures are generated from experience and insights documented in literature. It does therefore not claim to be exhaustive and furthermore evade the necessary effort to design and specify potential measures for the individual context of application. The assignment of distinct means for improvement relies consequently still on experience and the collaboration with component and machine tool manufacturers in order to realize saving potentials. The developed process routines empower however to revise means for improvement in a distinct preferential order.

Mastering the identified critical aspects of the energy performance management, a considerable improvement in the energy usage of machine tools can emanate from the implementation of the developed concept, which contributes to the initial motivation to bridge the energy efficiency gap.

6.3 Outlook on Future Research

Based on the review and the introduced restrictions for the application of the concept for energy performance management, consecutive fields of research can be formulated extending the scope of this work. These are briefly outlined below:

- The energy performance management is composed of several methods and tools, which involve the aggregation of measured quantities to indicator values as well the identification of efficient machine tools in single manual steps. Implementing the routines and activities of the energy performance management into a software-based environment provides an opportunity to ease the effort for the evaluation and improvement of energy efficiency. Extending the implementation towards a shared database of observed activities, the estimation of the efficient energy consumption function could be realized based on these collected activities enabling to specify a contemporary, distinct efficiency frontier. This unifying reference would consequently encourage the application of performance management also in small and medium sized enterprises, which are yet constrained by the effort the initially determine an efficient consumption function.
- The developed concept is exemplified for grinding machine tools as a representative of separating processes. These enable to relate the input energy demand

to the removed material volume as an output of the machining process. This provides an energy-oriented characterization based on the prevailing input–output relation facilitating the application of the methods for performance analysis. Transferring the energy performance management to forming processes opens up the challenge to define a functional reference, which enables to characterize distinctively the energy consumption in relation to the value creation.

- The energy performance management defines a technological system boundary based on experience and empirical observations. In order to enhance the accuracy of the performance management and avoid misleading comparisons of divergent technologies, it is essential to deduce clear and decisive criteria and factors, which provide an immanent characterization of a technological system boundary from an energy-oriented perspective. This distinction enables essentially to differentiate between technologies and assign machine tools to the adequate reference.
- The current scope for the introduced concept is limited to the usage of electrical energy to conduct the material removal. It does therefore not valuate the substitution of electrical energy with other energy forms. Additionally, machine tools require also compressed air and coolants to operate, which substantially exert an influence on the overall environmental performance of machining. Extending the energy performance management towards all resource flows, the determination of efficiency is challenged to define a unifying functional reference for the evaluation of environmental efficiency. It has to cope furthermore with a multitude of possible influencing quantities and variables amplifying the complexity of decision-making.

Appendix

The appendix provides an additional source of information with a supplementary view on the allocation of power demands, the experimental characterization of an energy consumption function for grinding machine tools, the determination of an efficient consumption function and the quantification of the technical efficiency in the industrial case study.

Appendix A: Allocation of Power Demands to Energy Flows

See Fig. A.1.

Appendix B: Design of Experiments for a Grinding Machine Tool

Determination of the specific energy demand based on the statistical characterization of the interrelations between process parameters and the energy demand (Tables A.1, A.2, A.3, Fig. A.2).

Appendix C: Linear Segments of the Enveloping Consumption Function

Calculation of the linear segments based on the linear interpolation between associated efficient activities (Table A.4).

A. Zein, *Transition Towards Energy Efficient Machine Tools*, Sustainable Production, Life Cycle Engineering and Management, DOI: 10.1007/978-3-642-32247-1,

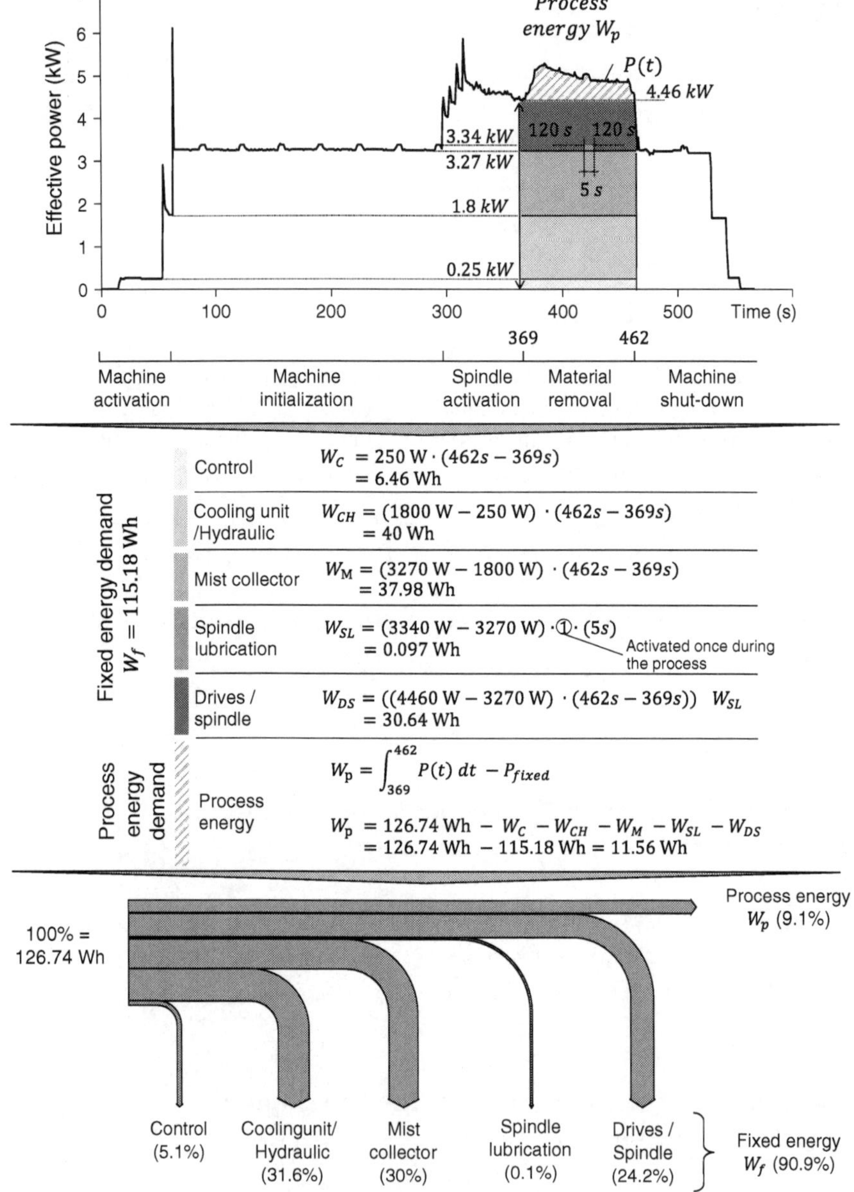

Fig. A.1 State-based assignment of power demands to components

Appendix D: Technical Efficiency of Grinding Machine Tools

Calculation of technical efficiency based on the comparison of the actual energy performance with the corresponding ideal energy performance (Table A.5).

Table A.1 Process setting and parameters

Parameter	Symbol	Unit	Specification
Process			Internal cylindrical grinding
Material			1.3505 62HRC 110 × 10 (mm)
Grinding wheel			B126 V M 8 VD 49 40 × 15 × 30 (mm)
Cutting fluid			Water based cutting fluid
Material removal rate	υ	mm^3/s	2–15
Cutting velocity	σ	m/s	40–60
Removed material volume	ρ	mm^3	500–2000

Table A.2 Set of experimental runs for the design of experiments

Experiment	Cutting Velocity σ (m/s)	Material removal rate υ (mm^3/s)	Removed material volume $\rho(mm^3)$
1	50.00	15	50.00
2	50.00	8.5	125.00
3	50.00	8.5	125.00
4	50.00	2	50.00
5	60.00	8.5	50.00
6	50.00	2	200.00
7	40.00	2	125.00
8	50.00	8.5	125.00
9	50.00	8.5	125.00
10	40.00	15	125.00
11	50.00	805	125.00
12	60.00	2	125.00
13	40.00	8.5	200.00
14	50.00	15	200.00
15	40.00	8.5	50.00
16	60.00	15	125.00
17	60.00	8.5	200.00

Table A.3 DOE model summary statistics

Source	Standard deviation	R-squared	Adjusted R-squared	Predicted R-squared
Linear	2.49	0.7761	0.7245	0.5607
2FI	1.98	0.8917	0.8268	0.5654
Quadratic	1.2	0.9719	0.9359	0.8242
Cubic	1.31	0.9811	0.9243	–

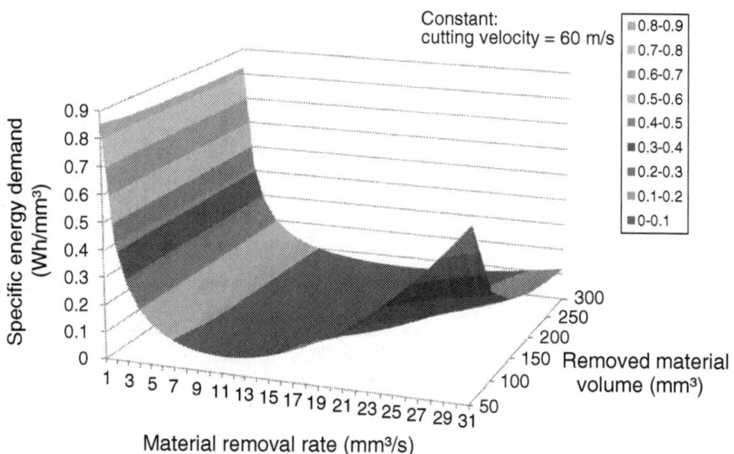

Fig. A.2 Dependence of the specific energy demand on the material removal rate and removed material volumeremoved material volume

Table A.4 Calculation of linear segments enveloping the efficient consumption function

Activities (m: gradient, c: constant)		Linear equation of enveloping segments SEC(MRR) = m MRR + c	Application range
1	$0.7 = m \cdot 2.64 + c$	SEC(MRR) = −0.7205 MRR + 2.5998	$2.64 \leq MRR \leq 2.82$
2	$0.56 = m \cdot 2.82 + c$		
2	$0.56 = m \cdot 2.82 + c$	SEC(MRR) = −0.2238 MRR + 1.197	$2.82 \leq MRR \leq 3.71$
5	$0.37 = m \cdot 3.71 + c$		
5	$0.37 = m \cdot 3.71 + c$	SEC(MRR) = −0.0383 MRR + 0.5093	$3.71 \leq MRR \leq 10.55$
7	$0.11 = m \cdot 10.55 + c$		
7	$0.11 = m \cdot 10.55 + c$	SEC(MRR) = −0.0038 MRR + 0.1453	$10.55 \leq \lambda_{MRR} \leq 14.43$
12	$0.09 = m \cdot 14.43 + c$		
12	$0.09 = m \cdot 14.43 + c$	SEC(MRR) = −0.0024 MRR + 0.1254	$14.43 \leq MRR \leq 28.03$
20	$0.06 = m \cdot 28.03 + c$		
20	$0.06 = m \cdot 28.03 + c$	SEC(MRR) = −0.0018 MRR + 0.1068	$28.03 \leq MRR \leq 41.65$
27	$0.03 = m \cdot 41.65 + c$		
27	$0.03 = m \cdot 41.65 + c$	SEC(MRR) = −0.0006 MRR + 0.0083	$41.65 \leq MRR \leq 50.29$
30	$0.04 = m \cdot 50.29 + c$		

Table A.5 Calculation of the technical efficiencies for the investigated grinding machine tools

			Linear seqments of efficient consumption function				
			SEC (MRR) = m. MRR + c				
Machine tool	Material removal rate (mm^3/s)	Actual SEC (Wh/mm^3)	Gradient m	Constant C	Ideal SEC (Wh/mm^3)	Technical efficiency (Wh/mm^3)	Relative technical efficiency (%)
1	2.64	0.70	−0.7205	2.5998	0.70	0.0000	0.0000
2	2.82	0.56	−0.7205	2.5998	0.56	0.0000	0.0000
3	3.04	0.80	−0.2238	1.1970	0.52	0.2809	0.3524
4	3.30	0.47	−0.2238	1.1970	0.46	0.0085	0.0182
5	3.71	0.37	−0.2238	1.1970	0.37	0.0000	0.0000
6	8.72	0.43	−0.0383	0.5093	0.18	0.2545	0.5921
7	10.55	0.11	−0.0383	0.5093	0.11	0.0000	0.0000
8	11.04	0.24	0.0038	0.1453	0.10	0.1400	0.5754
9	11.26	0.16	0.0038	0.1453	0.10	0.0560	0.3532
10	12.29	0.20	0.0038	0.1453	0.10	0.1019	0.5083
11	13.92	0.11	0.0038	0.1453	0.09	0.0144	0.1345
12	14.43	0.09	0.0038	0.1453	0.09	0.0000	0.0000
13	14.72	0.13	−0.0024	0.1254	0.09	0.0408	0.3116
14	19.68	0.18	−0.0024	0.1254	0.08	0.0978	0.5558
15	20.72	0.12	−0.0024	0.1254	0.08	0.0421	0.3572
16	22.40	0.09	−0.0024	0.1254	0.07	0.0187	0.2069
17	23.83	0.13	−0.0024	0.1254	0.07	0.0639	0.4838
18	24.88	0.09	−0.0024	0.1254	0.07	0.0260	0.2838
19	26.23	0.07	−0.0024	0.1254	0.06	0.0062	0.0902
20	28.03	0.06	−0.0024	0.1254	0.06	0.0000	0.0000
21	28.86	0.10	−0.0018	0.1068	0.05	0.0497	0.4756
22	31.49	0.10	−0.0018	0.1068	0.05	0.0497	0.4979
23	32.05	0.11	−0.0018	0.1068	0.05	0.0581	0.5417
24	33.22	0.08	−0.0018	0.1068	0.05	0.0363	0.4358
25	37.01	0.04	−0.0018	0.1068	0.04	0.0043	0.0973
26	38.34	0.06	−0.0018	0.1068	0.04	0.0230	0.3782
27	41.65	0.03	−0.0018	0.1068	0.03	0.0011	0.0320
28	44.90	0.09	0.0006	0.0083	0.04	0.0551	0.6097
29	45.85	0.07	0.0006	0.0083	0.04	0.0298	0.4543
30	50.29	0.04	0.0006	0.0083	0.04	0.0000	0.0000
	Measured activities		Calculated activities				

References

Abele E, Dervisopoulos M, Kuhrke B (2009) Bedeutung und Anwendung von Lebenszyklusanalysen bei Werkzeugmaschinen. In: Schweiger S (ed) Lebenszykluskosten optimieren. Paradigmenwechsel für Anbieter und Nutzer von Investitionsgütern, Gabler, pp 51–80

Abele E, Sielaff T, Schiffler A, Rothenbücher S (2011) Analyzing energy consumption of machine tool spindle units and identification of potential for improvements of efficiency. In: Hesselbach J, Herrmann C (eds) Proceedings of the 18th CIRP international conference on life cycle engineering glocalized solutions for sustainability in manufacturing. Springer, Berlin, pp 280–285

Adam D (2001) Produktions-management, 9th edn. Gabler, Wiesbaden

Aigner D, Lovell C, Schmidt P (1977) Formulation and estimation of stochastic frontier production function models. J Econom 6:21–37

Albach H (1962) Produktionsplanung auf der Grundlage technischer Verbrauchsfunktionen. In: Brandt L (ed) Natur-, Ingenieur- und Gesellschaftswissenschaften. Westdeutscher Verlag, Köln and Opladen, pp 45–96

Albach H (1980) Average and best-practice production functions in German Industry. J Ind Econ 29:55–70. http://www.jstor.org/pss/2097880

Allen K, Dyckhoff H (2002) Messung ökologischer Effizienz mittels Data Envelopment Analysis. Dissertation, 1st edn. Dt. Universitäts-Verlag, Wiesbaden

Almeida AT de, Ferreira FJTE, Fong J, Fonseca P (2008) EUP Lot 11 Motors. Final report. http://www.eup-network.de/fileadmin/user_upload/Produktgruppen/Lots/Final_Documents/Lot11_Motors_FinalReport.pdf. Accessed Aug 15 2011

Amaratunga D, Baldry D (2002) Moving from performance measurement to performance management. Facilities 20:217–223

Anderberg SE, Kara S, Beno T (2010) Impact of energy efficiency on computer numerically controlled machining. Proc Inst Mechan Eng B: J Eng Manuf 224:531–541

Andor M (2009) Die Bestimmung von individuellen Effizienzvorgaben—Alternativen zum Best-of-Four-Verfahren. Zeitschrift für Energiewirtschaft 33:195–204

Antony J (2010) Design of experiments for engineers and scientists, Reprinted. Butterworth-Heinemann, Amsterdam

Armstrong JS, Shapiro AC (1974) Analyzing quantitative models. J Mark 38:61–66

Astakhow VP, Outeiro JC (2008) Metal cutting mechanics, finite element modelling. In: Davim JP (ed) Machining. fundamentals and recent advances. Springer, London, pp 1–27. http://dx.doi.org/10.1007/978-1-84800-213-5

Atkins PW, de Paula J (2010) Atkins' physical chemistry, 9th edn. Oxford University Press, Oxford

A. Zein, *Transition Towards Energy Efficient Machine Tools*, Sustainable Production, Life Cycle Engineering and Management, DOI: 10.1007/978-3-642-32247-1,

Avram OI, Stroud I, Xirouchakis P (2011) A multi-criteria decision method for sustainability assessment of the use phase of machine tool systems. Int J Adv Manuf Technol 53:811–828

Avram OI, Xirouchakis P (2011) Evaluating the use phase energy requirements of a machine tool system. J Clean Prod 19:699–711

Baetge J (1974). Betriebswirtschaftliche Systemtheorie. Regelungstheoretische Planungs-Überwachungsmodelle für Produktion, Lagerung und Absatz. Westdt. Verlag, Opladen

Bartz WJ (1988) Energieeinsparung durch tribologische Maßnahmen. expert-Verlag, Ehningen bei Böblingen

Baumgartner W, Ebert O, Weber F (2006) Der Energieverbrauch der Industrie, 1990–2035. Ergebnisse der Szenarien I bis IV und der zugehörigen Sensitivitäten BIP hoch, Preise hoch und Klima wärmer. http://www.bfe.admin.ch/php/modules/enet/streamfile.php?file=000000010447.pdf

Beer S (1966) Decision and control the meaning of operational research and management cybernetics. Wiley, London

Behrendt T, Zein A, Min S (2012) Development of an energy consumption monitoring procedure for machine tools. CIRP Ann—Manufacturing Technology 61

Bellgran M, Säfsten K (2010) Production development. Design and operation of production systems. Springer, London. http://dx.doi.org/10.1007/978-1-84882-495-9

Berndt ER (1983) Electrification, energy quality, and productivity growth in U.S. manufacturing, Cambridge. http://openlibrary.org/books/OL14006001M/Electrification_energy_quality_and_productivity_growth_in_U.S._manufacturing

Binding HJ (1988) Grundlagen zur systematischen Reduzierung des Energie- und Materialeinsatzes. Dissertation, Aachen

Blackburn WR (2008) The sustainability handbook. the complete management guide to achieving social, economic and environmental responsibility, Reprinted. Earthscan, London

Bode H, Brecher C, Esser M, Hennes N, Klein WH, Prust D, Schreiber M, Wille H, Witt S, Würz T (2008) Mit intelligenten Lösungen zu optimierter Wertschöpfung. Entwicklungen im Werkzeugmaschinenbau. In: Brecher C, Klocke F (eds) Wettbewerbsfaktor Produktionstechnik. Aachener Perspektiven; Tagungsband; AWK, Aachener Werkzeugmaschinen-Kolloquium. Apprimus Verlag, Aachen, pp 171–201

Bode H-O (2007) Einfluss einer energieeffizienten Produktion auf Planungs- und Produktprämissen am Beispiel der Motorenfertigung. In: Proceedings of the XII. Internationales Produktionstechnisches Kolloquium, Berlin, pp 299–305

Bogetoft P, Otto L (2011). Benchmarking with DEA, SFA, and R. Springer Science + Business Media LLC, New York, NY. http://ebooks.ciando.com/book/index.cfm/bok_id/203563

Boulding KE (1956) General systems theory—the skeleton of science. Manage Sci 2:197–208. http://www.jstor.org/stable/2627132

Bourne M, Neely A, Mills J, Platts K (2003) Implementing performance measurement systems: a literature review. Int J Bus Perform Manage 5:1–24

Boyd G, Dutrow E, Tunnessen W (2008) The evolution of the ENERGY STAR® energy performance indicator for benchmarking industrial plant manufacturing energy use. J Clean Prod 16:709–715

Branham MS, Gutowski TG, Jones A, Sekulic DP (2008) A thermodynamic framework for analyzing and improving manufacturing processes. In: Proceedings of the IEEE international symposium on electronics and the environment. IEEE, pp 1–6

Braz RGF, Scavarda LF, Martins RA (2011) Reviewing and improving performance measurement systems: an action research. Int J Prod Econ 133:751–760

Brecher C, Bäumler S (2011) Zukunftsthemen im Werkzeugmaschinenbau. Hersteller müssen Standortvorteile in innovative Lösungen umsetzen. presentation slides. http://www.mav-online.de/c/document_library/get_file?uuid=a729ad48-6800-42e8-91b6-363f3dbe8cef&groupId=32571331. Accessed Sept 26 2011

Brecher C, Herfs W, Heyers C, Klein W, Triebs J, Beck E, Dorn T (2010) Ressourceneffizienz von Werkzeugmaschinen im Fokus der Forschung. Effizienzsteigerung durch Optimierung der Technologien zum Komponentenbetrieb. wt Werkstattstechnik online, pp 559–564. http://www.technikwissen.de/wt/article.php?data%5Barticle_id%5D=55628

Brecher C, Weck M (2005) Werkzeugmaschinen. Maschinenarten und Anwendungsbereiche. Springer, Berlin. http://dx.doi.org/10.1007/3-540-28085-5

Brecher C, Weck M (2006) Werkzeugmaschinen. Mechatronische Systeme, Vorschubantriebe, Prozessdiagnose, 6th edn. Springer, Berlin

Bruggink J (2011) Dealing with Doom: tackling the triple challenge of energy scarcity, climate change and global inequity. In: Jaeger CC, Tàbara JD, Jaeger J (eds) European research on sustainable development. Springer, Berlin, pp 59–71

Bunse K, Sachs J, Vodicka M (2010) Evaluating energy efficiency improvements in manufacturing processes. In: Vallespir B, Alix T (eds) IFIP advances in information and communication technology. Springer, Berlin, pp 19–26

Bunse K, Vodicka M, Schönsleben P, Brülhart M, Ernst FO (2011) Integrating energy efficiency performance in production management—gap analysis between industrial needs and scientific literature. J Clean Prod 19:667–679

Burger A (2008) Produktivität und Effizienz in Banken: Terminologie, Methoden und Status quo 92, Frankfurt a. M. https://www.econstor.eu/dspace/bitstream/10419/27858/1/577675443.PDF. Accessed Jan 17 2012

Cantner U, Hanusch H, Krüger J (2007) Produktivitäts- und Effizienzanalyse Der nichtparametrische Ansatz. Springer, Berlin

Capehart BL, Turner WC, Kennedy WJ (2012) Guide to energy management, 7th edn. Fairmont Press; Distributed by Taylor & Francis, Lilburn

Centre for Business Performance (2004) Literature review on performance measurement and management. The IDeA and audit commission performance management, measurement and information (PMMI) Project. http://www.idea.gov.uk/idk/aio/306299. Accessed Oct 31 2011

CIRP (1990) CIRP Annals 1990—Manufacturing technology. Nomenclature and definitions for manufacturing systems, 39th edn

Coelli TJ, Battese GE, O'Donnell CJ, Prasada Rao DS (2005) An introduction to efficiency and productivity analysis, 2nd edn. Springer Science+Business Media Inc, Boston, MA. http://dx.doi.org/10.1007/b136381

Cooper WW, Seiford LM, Tone K (2007) Data envelopment analysis. A comprehensive text with models, applications, references and DEA-solver software, 2nd edn. Springer Science+Business Media LLC, Boston, MA. http://ebooks.ciando.com/book/index.cfm/bok_id/254445

Dahmus JB, Gutowski TG (2004) An environmental analysis of machining. In: Proceedings of the 2004 ASME international mechanical engineering congress and RD&D Exposition, pp 643–652

Daraio C, Simar L (2007) Advanced robust and nonparametric methods in efficiency analysis. Methodology and applications. Springer, New York

DeCanio SJ (1993) Barriers within firms to energy-efficient investments. Energy Policy 21:906–914

Degner W (1986) Rationeller Energieeinsatz in der Teilefertigung, 1st edn. Verlag Technik, Berlin

DeGroff A, Schooley M, Chapel T, Poister TH (2010) Challenges and strategies in applying performance measurement to federal public health programs. Eval Progr Plan 33:365–372

Dervisopoulus M, Schatka A, Torney M, Warwela M (2006) Chancen und Risiken von Life Cycle Costing. Industriemanagement, pp 71–75

Desmira N, Narita H, Fujimoto H (2010) A minimization of environmental burden of high-speed milling. In: Shirase K, Aoyagi S (eds) Service robotics and mechatronics. Springer, London, pp 367–372

Deutsches Institut für Normung e.V (1985) Werkzeugmaschinen für die Metallbearbeitung. Beuth publisher, Berlin (69651)

Deutsches Institut für Normung e.V (2003) Fertigungsverfahren. Beuth publisher, Berlin (8580)

Deutsches Institut für Normung e.V (2006) Preparation of Documents used in Electrotechnology—Part 1: Rules (61082-1)

Devoldere T, Dewulf W, Deprez W, Duflou J (2008) Energy related life cycle impact and cost reduction opportunities in machine design. The laser cutting case. In: Kaebernick H (ed)

Applying life cycle knowledge to engineering solutions. LCE 2008, 15th CIRP International Conference on Life Cycle Engineering, Sydney, pp 412–419

Devoldere T, Dewulf W, Deprez W, Willems B, Duflou JR (2007) Improvement potential for energy consumption in discrete part production machines. In: Takata S, Umeda Y (eds) Proceedings of the 14th CIRP conference on life cycle engineering. Advances in life cycle engineering for sustainable manufacturing businesses. Springer, London, pp 311–316

Dhavale DG (1996) Performance measurement. Problems with existing manufacturing performance measures. J Cost Manag 9:50–55

Diaz N, Choi S, Helu M, Chen Y, Jayanathan S, Yasui Y, Kong D, Pavanaskar S, Dornfeld D (2010a) Machine tool design and operation strategies for green manufacturing. In: Proceedings of 4th CIRP international conference on high performance cutting, pp 271–276. http://www.me.berkeley.edu/~sushrut/papers/HPC2010-D08.pdf

Diaz N, Helu M, Jayanathan S, Yifen Chen, Horvath A, Dornfeld D (2010b) Environmental analysis of milling achine tool use in various manufacturing environments. In: Proceedings of the 2010 IEEE international symposium on sustainable systems & technology (ISSST). IEEE, Piscataway, NJ, pp 1–6

Diaz N, Redelsheimer E, Dornfeld D (2011) Energy consumption characterization and reduction strategies for milling machine tool use. In: Hesselbach J, Herrmann C (eds) Proceedings of the 18th CIRP international conference on life cycle engineering glocalized solutions for sustainability in manufacturing. Springer, Berlin, pp 263–267

Diekmann J, Eichhammer W, Neubert A, Rieke H, Schlomann B, Ziesing H-J (1999) Energie-Effizienz-Indikatoren. Statistische Grundlagen, theoretische Fundierung und Orientierungsbasis für die politische Praxis. Physica-Verlag, Heidelberg

Dietmair A, Verl A (2008) Energy consumption modeling and optimization for production machines. In: Institute of electrical and electronics engineers; international conference on sustainable energy technologies, IEEE international conference on sustainable energy technologies. IEEE, Piscataway, NJ, 574–579

Dietmair A, Verl A (2009a) A generic energy consumption model for decision making and energy efficiency optimisation in manufacturing. Int J Sustain Eng 2:123–133

Dietmair A, Verl A (2009b) Energy consumption forecasting and optimisation for tool machines. Mod Mach (MM) Sci J 63–67. http://www.mmscience.eu/archives/MM_Science_20090305.pdf

Dietmair A, Verl A (2010) Energy consumption assessment and optimisation in the design and use phase of machine tools. In: Proceedings of 17th CIRP international conference on life cycle engineering. Fundamental theories, sustainable manufacturing: application technologies and future development, pp 116–121

Dietmair A, Zulaika J, Sulitka M, Bustillo A, Verl A (2010) Lifecycle impact reduction and energy savings through light weight eco-design of machine tools. In: Proceedings of 17th CIRP international conference on life cycle engineering. Fundamental theories, sustainable manufacturing: application technologies and future development, pp 105–110

Dincer I, Rosen MA (2007) Exergy. Energy, environment and sustainable development. Elsevier, Amsterdam

DMG (2012). DMG energysave. product description. http://www.dmg.com/query/internet/v3/igpdf.nsf/745531e31ce887abc12576d4004a3bd8/$file/mailing_energysave_scan.pdf. Accessed Mar 27 2012

Draganescu F, Gheorghe M, Doicin CV (2003) Models of machine tool efficiency and specific consumed energy. J Mater Process Technol 141:9–15

Duflou J, Kellens K, Dewulf W (2011) Unit process impact assessment for discrete part manufacturing: a state of the art. CIRP J Manuf Sci Technol 4:129–135

Dyckhoff H (1994) Betriebliche Produktion. Theoretische Grundlagen einer umweltorientierten Produktionswirtschaft, 2 rev edn. Springer, Berlin

Dyckhoff H (2006) Produktionstheorie. Grundzüge industrieller Produktionswirtschaft, 5 rev edn. Springer, Berlin

Dyckhoff H, Souren R (2008) Nachhaltige Unternehmensführung. Grundzüge industriellen Umweltmanagements. Springer, Berlin

Dyckhoff H, Spengler TS (2010) Produktionswirtschaft. Eine Einführung, 3 rev edn. Springer, Berlin

Dyer SA (2001) Survey of instrumentation and measurement. Wiley, New York

Eckebrecht J (2000) Umweltverträgliche Gestaltung von spanenden Fertigungsprozssen. Shaker, Aachen

Eisele C, Schrems S, Abele E (2011) Energy-efficient machine tools through simulation in the design process. In: Hesselbach J, Herrmann C (eds) Proceedings of the 18th CIRP international conference on life cycle engineering. Glocalized solutions for sustainability in manufacturing. Springer, Berlin, pp 258–262

Engelmann J (2009) Methoden und Werkzeuge zur Planung und Gestaltung energieeffizienter Fabriken. Dissertation, Chemnitz

Engelmann J, Strauch J, Müller E (2008) Energieeffizienz als Planungsprämisse. Ressourcen- und Kostenoptimierung durch eine energieeffizienzorientierte Fabrikplanung. http://www.competence-site.de/downloads/46/85/i_file_805/energieeffizienz_als_planungspraemisse_100608.pdf. Accessed Aug 16 2011

Erdmann M-K (2002) Supply chain performance measurement. Dissertation, 1st edn. Eul, Lohmar, Dortmund

Erlach K, Westkämper E (eds) (2009) Energiewertstrom Der Weg zur energieeffizienten Fabrik. Fraunhofer Publishing, Stuttgart

European Commission (2008a) Action plan for sustainable consumption and production and sustainable industrial policy. Communication from the commission to the European Parliament, the Council, the European Economic and Social Committee and the Committee of the Regions. http://eur-lex.europa.eu/LexUriServ/LexUriServ.do?uri=COM:2008:0397:FIN:en:PDF

European Commission (2008b) Establishment of the working plan for 2009–2011 under the Ecodesign Directive. Communication from the Commission to the Council and the European Parliament. http://eur-lex.europa.eu/LexUriServ/LexUriServ.do?uri = COM:2008:0660:FIN:EN:PDF

European Commission (2008c) EU action against climate change. The EU emissions trading scheme, 2009th ed. Office for Official Publications of the European Communities, Luxembourg

European Commission (2009) Directive 2005/32/EC of the European Parliament and of the council with regard to ecodesign requirements for electric motors. Commission Regulation (EC) No 640/2009 of 22 July 2009. http://eur-lex.europa.eu/LexUriServ/LexUriServ.do?uri=OJ:L:2009:191:0026:0034:EN:PDF

European Commission (2010) Energy 2020. A strategy for competitive, sustainable and secure energy. http://eur-lex.europa.eu/LexUriServ/LexUriServ.do?uri = COM:2010:0639:FIN:EN:PDF

European Parliament, Council (2006) Directive 2006/42/EC of the European Parliament and of the Council of 17 May 2006 on machinery, and amending Directive 95/16/EC (recast). Official Journal of the European Union. http://eur-lex.europa.eu/LexUriServ/site/en/oj/2006/l_157/l_15720060609en00240086.pdf

European Parliament, Council (2009). Directive 2009/125/EC of the European Parliament and of the Council of 21 October 2009 establishing a framework for the setting of ecodesign requirements for energy-related products. http://eur-lex.europa.eu/LexUriServ/LexUriServ.do?uri=OJ:L:2009:285:0010:0035:EN:PDF

Eversheim W, Schuh G (2005) Integrierte Produkt- und Prozessgestaltung. Springer, Berlin. http://site.ebrary.com/lib/alltitles/docDetail.action?docID=10182951

Fandel G (2010) Produktions- und Kostentheorie, 8th edn. Springer, Berlin

Fang K, Uhan N, Zhao F, Sutherland JW (2011) A new shop scheduling approach in support of sustainable manufacturing. In: Hesselbach J, Herrmann C (eds) Proceedings of the 18th CIRP

International Conference on Life Cycle Engineering. Glocalized solutions for sustainability in manufacturing. Springer, Berlin, pp 305–310

Färe R, Grosskopf S, Lovell CAK (1985) The measurement of efficiency of production. Kluwer-Nijhoff, Boston

Federal Statistical Office (2011) Statistisches Jahrbuch für die Bundesrepublik Deutschland. editorial deadline: 1 Aug 2011, Wiesbaden. http://www.destatis.de/jetspeed/portal/cms/Sites/destatis/SharedContent/Oeffentlich/B3/Publikation/Jahrbuch/StatistischesJahrbuch,property=file.pdf

Ferreira A, Otley D (2009) The design and use of performance management systems: an extended framework for analysis. Manag Acc Res 20:263–282

Folan P, Browne J (2005) A review of performance measurement: towards performance management. Comput Ind 56:663–680

Fridman A (2012) The quality of measurements. A metrological reference. Springer Science+Business Media LLC, New York

Fruehan RJ, Fortini O, Paxton HW, Brindle R (2000) Theoretical minimum energies to produce steel for selected conditions. http://www1.eere.energy.gov/industry/steel/pdfs/theoretical_minimum_energies.pdf. Accessed Aug 16 2011

Fanuc GE (2008) The environmental and economic advantages of energy-efficient motors. http://leadwise.mediadroit.com/files/2928energy%20saving_wp_gft688.pdf. Accessed Sept 28 2011

Gahrmann A, Hempfling R, Sietz M (1993) Bewertung betrieblicher Umweltschutzmaßnahmen. Ökologische Wirksamkeit und ökonomische Effizienz, Blottner

Galitsky C, Worrell E (2008) Energy efficiency improvement and cost saving opportunities for the vehicle assembly industry. An ENERGY STAR Guide for Energy and Plant Managers. http://ies.lbl.gov/iespubs/energystar/vehicleassembly.pdf

German Electrical and Electronic Manufacturers' Association (2010) Electric motors and variable speed drives. Standards and legal requirements for the energy efficiency of low-voltage three-phase motors. http://www.zvei.org/fileadmin/user_upload/Fachverbaende/Automation_Antriebe/energieffizienz/ZVEI_Electric_Motors_and_Variable_Speed_Drives_2nd_Edition-Internet-Version.pdf. Accessed Sept 7 2011

Gleason (2011) Gleason energy saver package. Product description. http://www.gleason-pfauter.ch/service/Flyer%20Eco_Packet_EN_2011_05_27.pdf. Accessed Mar 23 2012

Gomez P, Malik F, Oeller K-H (1975) Systemmethodik. Grundlagen einer Methodik zur Erforschung und Gestaltung komplexer soziotechnischer Systeme. Verlag Paul Haupt, Bern

Goossens Y, Mäkipää A, Schepelmann P, van de Sand I, Kuhndtand M, Herrndorf M (2007) Alternative progress indicators to Gross Domestic Product (GDP) as a means towards sustainable development. Economic and Scientific Policy. Study for the European Parliament's Committee on the Environment, Public Health. http://www.beyond-gdp.eu/download/bgdp-bp-goossens.pdf

Greene WH (2008) The econometric approach to efficiency analysis. In: Fried HO, Lovell CAK, Schmidt SS (eds) The measurement of productive efficiency and productivity growth. Oxford University Press, Oxford, pp 92–250

Grüning M (2002) Performance-Measurement-Systeme. Messung und Steuerung von Unternehmensleistung. Dissertation, 1st ed. Dt. Universitäts-Verlag, Wiesbaden

Günther T (1991) Erfolg durch strategisches Controlling? Eine empirische Studie zum Stand des strategischen Controlling in deutschen Unternehmen und dessen Beitrag zu Unternehmenserfolg und -risiko. Dissertation. Vahlen, München

Gutowski TG, Dahmus JB, Thiriez A (2006) Electrical energy requirements for manufacturing processes. In: Proceedings of 13th CIRP international conference on life cycle engineering. Towards a closed loop economy, pp 623–627

Gutowski T, Dahmus JB, Thiriez A, Branham MS, Jones A (2007) A thermodynamic characterization of manufacturing processes. In: Computer Society; international association of electronics recyclers; IEEE international symposium on electronics & the environment; ISEE; international association of electronics recyclers (IAER) summit; Electronics recycling summit, Proceedings of the 2007 IEEE international symposium on electronics & the environment. 7–10 May 2007, Orlando, FL, USA; conference record; co-located with the

International Association of Electronics Recyclers (IAER) summit. IEEE Operations Center, Piscataway, NJ, 137–142

Haberfellner R (1999) Systems engineering. Methodik und Praxis, 10 rev. edn. Verlag Industrielle Organisation, Zürich

Hagemann D (2011) Status of ISO/TC39/WG12. 2nd Stakeholdermeeting of the EC Product Group Study for ENTR Lot 5 Machine Tools. http://www.ecomachinetools.eu/typo/meetings.html?file=tl_files/pdf/Statusreport_ISOWG12_28032011.pdf. Accessed Mar 1 2012

Harvey LDD (2010) Energy and the new reality 1. Earthscan, London

Hatry HP, Wholey JS (2006) Performance measurement. Getting results, 2nd edn. Urban Institute Press, Washington

Hegener G (2010). Energieeffizienz beim Betrieb von Werkzeugmaschinen—Einsparpotentiale bei der Auswahl der Fertigungstechnologie. Fertigungstechnisches Kolloquium Stuttgart, pp 281–292

Heidenhain (2010). Aspects of energy efficiency in machine tools. http://www.heidenhain.co.jp/fileadmin/pdb/media/img/Energieeffizienz_WZM_en.pdf. Accessed Sept 28 2011

Herrmann C (2010) Ganzheitliches life cycle management. Nachhaltigkeit und Lebenszyklusorientierung in Unternehmen. Springer, Berlin. http://dx.doi.org/10.1007/978-3-642-01421-5

Herrmann C, Bergmann L, Thiede S, Zein A (2007) Energy labels for production machines. An approach to facilitate energy efficiency in production systems. In: Proceedings of the 40th CIRP international seminar on manufacturing systems, Liverpool

Herrmann C, Bogdanski G, Zein A (2010) Industrial smart metering—application of information technology systems to improve energy efficiency in manufacturing. In: Sihn W, Kuhlang P (eds) Proceedings of the 43rd CIRP international conference on manufacturing systems. Sustainable production and logistics in global networks. NWV, Wien, pp 244–251

Herrmann C, Suh S-H, Bogdanski G, Zein A, Cha J-M, Um J, Jeong S, Guzman A (2011a) Context-aware analysis approach to enhance industrial smart metering. In: Hesselbach J, Herrmann C (eds) Proceedings of the 18th CIRP international conference on life cycle engineering. Glocalized solutions for sustainability in manufacturing. Springer, Berlin, pp 323–328

Herrmann C, Thiede S (2009) Process chain simulation to foster energy efficiency in manufacturing. CIRP J Manuf Sci Technol 1:221–229

Herrmann C, Thiede S, Kara S, Hesselbach J (2011b) Energy oriented simulation of manufacturing systems—Concept and application. CIRP Ann—Manuf Technol 60:45–48

Herrmann C, Thiede S, Zein A, Ihlenfeldt S, Blau P (2009) Energy efficiency of machine tools - extending the perspective. In: Proceedings of the 42nd CIRP conference on manufacturing systems. Sustainable Development of Manufacturing Systems, Grenoble

Hilgers D (2008) Performance-management. Leistungserfassung und Leistungssteuerung in Unternehmen und öffentlichen Verwaltungen. Dissertation, 1st ed. Gabler, Wiesbaden

Hill AV (2011) The encyclopedia of operations management. A field manual and glossary of operations management terms and concepts, 1st ed., 1. print. FT Press, Upper Saddle River, NJ

Hon K (2005) Performance and evaluation of manufacturing systems. CIRP Ann—Manuf Technol 54:139–154

Huber J (2000) Towards industrial ecology: sustainable development as a concept of ecological modernization. J Environ Policy Plan 2:69–285. http://dx.doi.org/10.1002/1522-7200(200010/12)2:4<269:AID-JEPP58>3.0.CO;2-U

International Atomic Energy Agency, United Nations Department of Economic and Social Affairs, International Energy Agency, Eurostat, European Environment Agency (2005) Energy indicators for sustainable development. Guidelines and methodologies, Vienna. http://www-pub.iaea.org/MTCD/publications/PDF/Pub1222_web.pdf

International Energy Agency (2006) Energy technology perspectives. In support of the G8 plan of action. Scenarios and strategies to 2050

International Energy Agency (2007) Mind the gap. Quantifying principal-agent problems in energy efficiency; in support of the G8 plan of action. International Energy Agency, Head of Communication and Information, Paris. http://www.iea.org/textbase/nppdf/free/2007/mind_the_gap.pdf

International Energy Agency (2009) World Energy Outlook, Paris

International Standard Organization (2006) Environmental management—Life cycle assessment—Principles and framework (ISO 14040:2006)

International Standard Organization (2011) Energy management systems—Requirements with guidance for use (ISO 50001:2011)

Jaffe A (1994) The energy-efficiency gap. What does it mean? Energy Policy 22:804–810

Japanese Standards Association (2010) Machine tools—Test methods for electric power consumption—Part 1: Machining Centres (TS B 0024-1:2010)

Jasch C, Tukker A (2009) Environmental and material flow cost accounting. Principles and procedures. Springer, Netherlands

Johnston R, Fitzgerald L, Markou E, Brignall S (2001) Target setting for evolutionary and revolutionary process change. Int J Operations & Prod Manag 21:1387–1403

Kalpakjian S, Schmid SR (2001) Manufacturing engineering and technology, 4th edn. Prentice Hall, Upper Saddle River

Kara S, Bogdanski G, Li W (2011) Electricity metering and monitoring in manufacturing systems. In: Hesselbach J, Herrmann C (eds) Proceedings of the 18th CIRP international conference on life cycle engineering. Glocalized solutions for sustainability in manufacturing. Springer, Berlin, pp 1–10

Kara S, Li W (2011) Unit process energy consumption models for material removal processes. CIRP Ann—Manuf Technol 60:37–40

Kellens K, Dewulf W, Deprez W, Yasa E, Duflou JR (2010) Environmental analysis of SLM and SLS manufacturing processes. In: Proceedings of 17th CIRP international conference on life cycle engineering. Fundamental theories, sustainable manufacturing: application technologies and future development

Kellens K, Dewulf W, Overcash M, Hauschild M, Duflou JR, Hauschild MZ (2011a) Methodology for systematic analysis and improvement of manufacturing unit process life-cycle inventory (UPLCI)—CO2PE! initiative (cooperative effort on process emissions in manufacturing). Part 1: Methodology description. International Journal of Life Cycle Assessment

Kellens K, Renaldi Dewulf W, Duflou JR (2011b) Preliminary environmental assessment of electrical discharge machining. In: Hesselbach J, Herrmann C (eds) Proceedings of the 18th CIRP international conference on life cycle engineering. Glocalized solutions for sustainability in manufacturing. Springer, Berlin, pp 377–382

Kemna R, van Elburg M, Li W, van Holsteijn R (2005) Methodology study eco-design of energy-using products. MEEUP methodology report. http://ec.europa.eu/enterprise/policies/sustainable-business/ecodesign/methodology/files/finalreport1_en.pdf. Accessed Nov 10 2011

Kienzle O (1952) Die Bestimmung von Kräften und Leistungen an spanenden Werkzeugen und Werkzeugmaschinen. In: VDI-Z, pp 299–305

Kistner K-P (1993) Produktions- und Kostentheorie, 2nd edn. Physica, Heidelberg

Kleine A, Dinkelbach W (2002) DEA-Effizienz. Entscheidungs- und produktionstheoretische Grundlagen der Data Envelopment Analysis. Habilitation thesis, 1st edn. Dt. Universitäts-Verlag, Wiesbaden

Koopmans CC, te Velde DW (2001) Bridging the energy efficiency gap: using bottom-up information in a top-down energy demand model. Energy Econ 23:57–75

Koopmans TC (1951) Analysis fo production as an efficient combination of activities. In: Koopmans TC, Alchian A, Dantzig GB, Georgescu-Roegen N, Samuelson PA, Tucker AW (eds) Activity analysis of production and allocation. Proceedings of a conference. Wiley, New York, pp 33–97. http://cowles.econ.yale.edu/P/cm/m13/m13-all.pdf

Koskela L (2000) An exploration towards a production theory and its application to construction. Technical Research Centre of Finland, Espoo

Krause O, Mertins K (2006) Performance management. Eine Stakeholder-Nutzen-orientierte und Geschäftsprozess-basierte Methode. Dissertation, 1st ed. Dt. Universitäts-Verlag, Wiesbaden

Kuhrke B, Schrems S, Eisele C, Abele E (2010) Methodology to assess the energy consumption of cutting machine tools. In: Proceedings of 17th CIRP international conference on life cycle engineering. Fundamental theories, sustainable manufacturing: application technologies and future development

Kumbhakar S, Lovell CAK (2003) Stochastic frontier analysis, 1st edn. Cambridge University Press, Cambridge

Larek R, Brinksmeier E, Meyer D, Pawletta T, Hagendorf O (2011a) A discrete-event simulation approach to predict power consumption in machining processes. Prod Eng Res Devel 5:575–579

Larek R, Brinksmeier E, Pawletta T, Hagendorf O (2011b) Model-based planning of resource efficient process chains using system entity structures. In: Proceedings of the 1st WGP-Jahreskongress

Li W, Kara S (2011) An empirical model for predicting energy consumption of manufacturing processes: a case of turning process. Proc Inst Mechan Eng B: J Eng Manuf 225:1636–1646

Li W, Winter M, Kara S, Herrmann C (2012) Eco-efficiency of manufacturing processes: a grinding case. CIRP Ann—Manuf Technol (in press)

Li W, Zein A, Kara S, Herrmann C (2011) An investigation into fixed energy consumption of machine tools. In: Hesselbach J, Herrmann C (eds) Proceedings of the 18th CIRP international conference on life cycle engineering. Glocalized solutions for sustainability in manufacturing. Springer, Berlin, pp 268–273

Lichiello P, Turnock BJ (1999) Guidebook for performance measurement, Seattle. http://www.turningpointprogram.org/toolkit/pdf/pmc_guide.pdf

Lundberg K, Balfors B, Folkeson L (2009) Framework for environmental performance measurement in a Swedish public sector organization. J Clean Prod 17:1017–1024

Malik F (2011) Corporate policy and governance. How organizations self-organize. Campus, Frankfurt am Main

Meadows DH, Meadows DL, Randers J (2001) Die neuen Grenzen des Wachstums, 5th edn. Rowohlt, Reinbek bei Hamburg

Meadows DH, Randers J, Meadows DL (2005) The limits to growth. The 30-year update, Rev. ed. Earthscan, London

Mellerowicz K (1963) Unternehmenspolitik. Haufe, Freiburg im Breisgau

Metz B, Davidson OR, Bosch PR, Dave R, Meyer LA (2007) Climate change 2007. Mitigation of climate change. Contribution of working group III to the fourth assessment report of the intergovernmental panel on climate change. Cambridge University Press, Cambridge

Mock A (1986) Wirtschaftskybernetische Erfahrungen in der Wirtschaftspraxis. In: Witte T (ed) Systemforschung und Kybernetik für Wirtschaft und Gesellschaft. Wissenschaftliche Jahrestagung der Gesellschaft für Wirtschafts- und Sozialkybernetik am 11. und 12. Oktober 1985 in Osnabrück. Duncker & Humblot, Berlin

Mori M, Fujishima M, Inamasu Y, Oda Y (2011) A study on energy efficiency improvement for machine tools. CIRP Ann—Manuf Technol 60:145–148

Mouzon G (2008) Operational methods and models for minimization of energy consumption in a manufacturing environment. Dissertation. http://soar.wichita.edu/dspace/bitstream/handle/10057/1954/d08006.pdf?sequence=1

Müller E, Engelmann J, Löffler T, Strauch J (2009) Energieeffiziente Fabriken planen und betreiben, 1st edn. Springer, Berlin

Müller L (2001) Handbuch der Elektrizitätswirtschaft. Technische, wirtschaftliche und rechtliche Grundlagen, 2nd ed. Springer, Berlin. http://www.gbv.de/dms/faz-rez/F19990125ELEKT-100.pdf

Mulvaney D (2011) Green technology. An A-to-Z guide. SAGE, Los Angeles

Narita H, Kawamura H, Norihisa T, Chen L-y, Fujimoto H, Hasebe T (2006) Development of prediction system for environmental burden for machine tool operation. Jpn Soc Mechan Eng, Int J Ser C 49:1188–1195

Narita J, Chen LY, Kawamura H, Fujimoto H (2003) Evaluation system for machine tool operation with considering the global environment. In: Environmental informatics archives, pp 348–356

National Performance Review (1997) Serving the American Public: best practices in performance measurement. Benchmarking Study Report. http://govinfo.library.unt.edu/npr/library/papers/benchmrk/nprbook.pdf

National Research Council (1979) Measurement and interpretation of productivity. Panel to review productivity statistics. National Academy of Sciences, Washington

Neely A, Gregory M, Platts K (2005) Performance measurement system design: a literature review and research agenda. Int J Operations & Prod Manag 25:1228–1263

Neugebauer R, Blau P, Harzbecker C, Weidlich D (2008) Ressourceneffiziente Maschinen- und Prozessgestaltung. In: Neugebauer R (ed) Zerspanung in Grenzbereichen. Machining on the cutting edge. Verlag Wissenschaftliche Scripten, pp 49–67. http://www.worldcat.org/oclc/271648201

Neugebauer R, Wabner M, Rentzsch H, Ihlenfeldt S (2011) Structure principles of energy efficient machine tools. CIRP J Manuf Sci Technol 4:136–147

Neugebauer R, Wertheim R, Hochmuth C, Schmidt G, Dix M (2010) Modelling of energy and resource-efficient machining. In: Aoyama T, Takeuchi Y (eds) Proceedings of the 4th CIRP international conference on high performance cutting. Faculty of Science and Technology, Yokohama, pp 295–300

Nolte A, Oppel J (2008) Klimawandel—eine Herausforderung für die Wirtschaft. Handlungsoptionen für Industrieunternehmen in Deutschland, 1st edn. Diplomica Verlag, Hamburg

Nudurupati S, Arshad T, Turner T (2007) Performance measurement in the construction industry: an action case investigating manufacturing methodologies. Comput Ind 58:667–676

Olson W (2006) Systems thinking. Chapter 5. In: Abraham MA (ed) Sustainability science and engineering. Defining principles. Elsevier, Amsterdam, pp 91–112. http://www.sciencedirect.com/science/book/9780444517128

Organisation for Economic Co-operation and Development (1999) Environmental requirements for industrial permitting. OECD Publications, Paris

Ostertag K, Jochem E, Schleich J, Walz R, Kohlhaas M, Diekmann J, Ziesing H-J (2000) Energiesparen—Klimaschutz, der sich rechnet. Ökonomische Argumente in der Klimapolitik. Physica-Verlag, Heidelberg

Otten K, Debons A (1970) Towards a metascience of information: Informatology. J Am Soc Infor Sci 21:89–94

Parthier R (2010) Messtechnik. Grundlagen und Anwendungen der elektrischen Messtechnik für alle technischen Fachrichtungen und Wirtschaftsingenieure; mit 31 Tabellen, 5., erw. Vieweg+Teubner, Wiesbaden

Patterson MG (1996) What is energy efficiency? Energy Policy 24:377–390

Pears A (2004) Energy efficiency—Its potential. Some perspectives and experiences. background paper. http://www.naturaledgeproject.net/Documents/IEAENEFFICbackgroundpaperPearsFinal.pdf

Pfeifer T, Schmitt R (2010) Fertigungsmesstechnik, 3rd edn. Oldenbourg, München

Plapper V, Weck M (2001) Sensorless machine tool condition monitoring based on open NCs. In: Robotics and automation Society; Institute of electrical and electronics engineers; IEEE international conference on robotics and automation; IEEE ICRA 2001, Proceedings of the IEEE international conference on robotics and automation (ICRA). IEEE Operations Center, Piscataway, NJ, pp 3104–3108

Pohselt D (2011) Alte Maschinen brauchen oft weniger Energie als neue. Industriemagazin. http://www.industriemagazin.net/home/artikel/Fertigungstechnik/Alte_Maschinen_brauchen_oft_weniger_Energie_als_neue/aid/7247

PROFIBUS Nutzerorganisation e. V (2011) Assessing PROFIenergy‘s potential. Quantifying the energy saving possibilities of PI‘s PROFIenergy profile for PROFINET and assessing its deployment opportunities. PI White Paper

Prolima (2008) Best environmental practice manual. Environmental Product Lifecycle Management for building competitive machine tools. EU research project. http://www.maquinaherramienta.biz/prolima-eu/best-practice-manual.htm. Accessed Oct 1 2011

Rager M (2008) Energieorientierte Produktionsplanung. Analyse, Konzeption und Umsetzung. Betriebswirtschaftlicher Verlag Dr. Th. Gabler/GWV Fachverlage GmbH Weisbaden, Wiesbaden. http://dx.doi.org/10.1007/978-3-8350-5569-8

Rajemi MF (2010) Energy analysis in turning and milling. Dissertation. https://www. escholar. manchester.ac.uk/api/datastream?publicationPid=uk-ac-man-scw: 119698&datastreamId = FULL-TEXT.PDF

Ray SC (2004) Data envelopment analysis. Theory and techniques for economics and operations research. Cambridge University Press, Cambridge. http://site.ebrary.com/lib/academiccompletetitles/home.action

Reap J, Bras B (2008) Exploring the limits to sustainable energy consumption for organisms and devices. Int J Sustain Manuf 1:78–99

Renaldi, Kellens, K, Dewulf W, Duflou JR (2011) Exergy efficiency definitions for manufacturing processes. In: Hesselbach J, Herrmann C (eds) Proceedings of the 18th CIRP international conference on life cycle engineering. Glocalized solutions for sustainability in manufacturing. Springer, Berlin, pp 329–334

Rogalski S (2011) Flexibility measurement in production systems. Handling uncertainties in industrial production. Springer, Berlin. http://ebooks.ciando.com/book/index.cfm/bok_id/267256

Rowe WB (2009) Principles of modern grinding technology, 1st. William Andrew, Oxford

Schiefer E (2000) Ökologische Bilanzierung von Bauteilen für die Entwicklung umweltgerechter Produkte am Beispiel spanender Fertigungsverfahren. Dissertation. Shaker, Aachen, Darmstadt

Schieferdecker B (2006) Energiemanagement-tools. Anwendung im Industrieunternehmen. Springer, Berlin. http://dx.doi.org/10.1007/3-540-29481-3

Schischke K, Hohwieler E, Feitscher R, König J, Kreuschner S, Nissen NF, Wilpert P (2011a) Energy-using product group analysis. Lot 5: Machine tools and related machinery. Executive Summary—Version 2. http://www.ecomachinetools.eu/typo/reports.html?file=tl_files/pdf/EuP_LOT5_ExecutiveSummary_v05_280211.pdf. Accessed Sept 11 2011

Schischke K, Hohwieler E, Feitscher R, König J, Kreuschner S, Nissen NF, Wilpert P (2011b) Energy-using product group analysis. Lot 5: Machine tools and related machinery. Task 1 report. http://www.ecomachinetools.eu/typo/reports.html?file=tl_files/pdf/EuP_LOT5_Task1_Draft_28-02-11_v3.pdf. Accessed Sept 11 2011

Schischke K, Hohwieler, E, Feitscher R, König J, Kreuschner S, Wilpert P, Nissen NF (2012a) Energy-using product group analysis. Lot 5: Machine tools and related machinery. Draft Task 7 – Policy and impact analysis. http://www.google.de/url?sa=t&rct=j&q=energy%20label%20for%20machine%20tools&source=web&cd=8&ved=0CH0QFjAH&url=http%3A%2F%2Fwww.ecomachinetools.eu%2Ftypo%2Freports.html%3Ffile%3Dtl_files%2Fpdf%2FEuP_Lot5_Task7_March2012.pdf&ei=S-mYT4vpGq3c4QSKo-zEBg&usg=AFQjCNF5MT0cdPKQdLfJZobT06QF04BWVQ. Accessed April 26 2012

Schischke K, Hohwieler E, Feitscher R, König J, Nissen NF, Wilpert P, Kreuschner S (2012b). Energy-using product group analysis. Lot 5: machine tools and related machinery. Draft Task 5—Technical Analysis BAT and BNAT. http://www.ecomachinetools.eu/typo/reports.html?file=tl_files/pdf/EuP_LOT5_Task5_Draft_11-02-28_v10.pdf. Accessed Mar 29 2012

Schleich J (2009) Barriers to energy efficiency: a comparison across the German commercial and services sector. Ecol Econ 68:2150–2159

Schmid C (2004) Energieeffizienz in Unternehmen. Eine wissensbasierte Analyse von Einflussfaktoren und Instrumenten. Dissertation. vdf Hochschulverlag, Zürich

Schmitt R, Bittencourt JL, Bonefeld R (2011) Modelling machine tools for self-optimisation of energy consumption. In: Hesselbach J, Herrmann C (eds) Proceedings of the 18th CIRP international conference on life cycle engineering. Glocalized solutions for sustainability in manufacturing. Springer, Berlin, pp 253–257

Schultz A (2002) Methode zur integrierten ökologischen und ökonomischen Bewertung von Produktionsprozessen und -technologien. Dissertation, Magdeburg

Schulz H, Schiefer E (1998) Prozessführung und Energiebedarf bei spanenden Fertigungsverfahren. Eine Analyse der Zusammenhänge. ZWF 93:266–271

Schulz H, Schiefer E (1999) Methodology for the life cycle inventory of machined parts. Prod Eng Res Devel 6(2):121–124

Schweiger S (ed) (2009) Lebenszykluskosten optimieren. Paradigmenwechsel für Anbieter und Nutzer von Investitionsgütern. Gabler, Wiesbaden

Seefeldt F, Wünsch M, Matthes U, Baumgartner W, Michelsen C, Orsolya E-B, Leybold P, Herz T (2006) Potenziale für Energieeinsparung und Energieeffizienz im Lichte aktueller Preisentwicklungen. final project report. http://www.bmwi.de/BMWi/Redaktion/PDF/Publikationen/Studien/studie-prognos-energieeinsparung,property=pdf,bereich=bmwi,sprache=de,rwb=true.pdf. Accessed Sept 7 2011

Seidl J (2002) Business process performance—Modellbezogene Beurteilung und Ansätze zur Optimierung. HMD—Praxis Wirtschaftsinform 227

Seryak J, Kissock K (2005) Lean energy analysis: guiding industrial energy reduction efforts to the theoretical minimum energy use. Accessed Aug 10 2011

Shin SJ (2009) Development of a framework for green productivity enhancement and its application to machining system. Pohang University of Science and Technology. Dissertation

Siemens AG (2010). Saving energy with SIMATIC S7. PROFIenergy with ET200. Application description. http://support.automation.siemens.com/WW/llisapi.dll/csfetch/41986454/41986454_PROFIenergy_ET200S_DOKU_V10_en.pdf?func=cslib.csFetch&nodeid=45058092. Accessed Mar 26 2012

Siemens AG (2011). Sizer for Siemens drives. http://support.automation.siemens.com/WW/view/de/10804987/133100

Singh SK, Kundu SC, Singh S (1998) Ecosystem management. Mittal Publications, New Delhi

Sink DS, Tuttle TC (1995) Planning and measurement in your organization of the future, 4th edn. Industrial Engineering and Management Press, Norcross

Sonntag S, Kistner K-P (2004) Die Gutenberg-Produktionsfunktion. Eigenschaften und technische Fundierung. Dissertation, 1st edn. Dt. Universitäts-Verlag, Wiesbaden

Stasinopoulos P, Smith MH, Hargroves K, Desha C (2009) Whole system design. An integrated approach to sustainable engineering. Earthscan, London

Steiner R, Frischknecht R (2007) Metals processing and compressed air supply. Ecoinvent report. EMPA Dübendorf, Dübendorf. http://www.ecoinvent.ch/

Stivers BP, Covin TJ, Green Hall N, Smalt SW (1998) How nonfinancial performance measures are used. Manag Acc 44:46–49

Stoop PPM (1996) Performance management in manufacturing. A method for short term performance evaluation and diagnosis. Dissertation, Eindhoven. http://alexandria.tue.nl/extra3/proefschrift/PRF12A/9600624.pdf

Sturm A (2000) Performance measurement und environmental performance measurement, [Elektronische Ressource], Dresden. http://hsss.slub-dresden.de/pub2/dissertation/2001/wirtschaftswissenschaften/994768126734-5500/994768126734-5500.pdf

Takada M, Morris E, Rajan SC (2000) Sustainable energy strategies: materials for decision-makers. http://www.undp.org/energy/publications/2000/2000a.htm

Tam C (2008) Developing indicators of energy efficiency. In: OECD Organisation for Economic Co-operation and Development (ed) Measuring sustainable production. Proceedings of an OECD workshop on sustainable manufacturing production and competitiveness. OECD, Paris, pp 25–38

Tanaka K (2008) Assessment of energy efficiency performance measures in industry and their application for policy. Energy Policy 36:2887–2902

Terry GR (1968) Principles of management, 5th edn. Irwin, Homewood Ill

Thiede S (2012) Energy efficiency in manufacturing systems. Springer, Berlin

Thumann A (2010) Plant engineers and managers guide to energy conservation, 10th edn. Fairmont Press, Liburn

Tönshoff HK (1995) Werkzeugmaschinen. Grundlagen. Springer, Berlin

U.S. Energy Information Administration (2007) International energy outlook. http://www.iea.org/textbase/nppdf/free/2007/weo_2007.pdf

U.S. Energy Information Administration (2010) International energy outlook. http://www.eia.gov/oiaf/ieo/pdf/0484(2010).pdf

United Nations (1998) Kyoto protocol to the United Nations framework convention on climate change. http://unfccc.int/resource/docs/convkp/kpeng.pdf

Verein Deutscher Ingenieure e.V (1998) Energy consulting for industry and business (3922)

Verein Deutscher Ingenieure e.V (2003) Energiekenngrößen. Definitionen, Begriffe, Methodik=Energetic characteristics: definitions, terms, methodology. Beuth, Berlin

Verl A, Abele E, Heisel U, Dietmair A, Eberspächer P, Rahäuser R, Schrems S, Braun S (2011) Modular modeling of energy consumption for monitoring and control. In: Hesselbach J, Herrmann C (eds) Proceedings of the 18th CIRP international conference on life cycle engineering. Glocalized solutions for sustainability in manufacturing. Springer, Berlin, pp 341–346

Vijayaraghavan A, Dornfeld D (2010) Automated energy monitoring of machine tools. CIRP Ann—Manuf Technol 59:21–24

Volkswagen AG (2011a) Gemeinsame Umwelterklärung 2010/Joint Environmental Statement 2010. Volkswagen and Volkswagen Commercial Vehicles. http://www.volkswagen.de/content/medialib/vwd4/de/Volkswagen/Nachhaltigkeit/service/download/umwelterklaerungen0/Gemeinsame_Umwelterklaerung_2010/_jcr_content/renditions/rendition.file/uwe_gesamt_2010.pdf. Accessed Sept 21 2011

Volkswagen AG (2011b) Umwelterklärung 2011/Environmental Statement 2011. Kassel. http://www.volkswagen.de/content/medialib/vwd4/de/Volkswagen/Nachhaltigkeit/service/download/umwelterklaerungen0/Kassel_2011/_jcr_content/renditions/rendition.file/kassel2011.pdf. Accessed April 1 2012

Waggoner DB, Neely AD., Kennerley MP (1999) The forces that shape organisational performance measurement systems: an interdisciplinary review. Int J Prod Econ 60–61:53–60

Wanke A, Trenz S (2001) Energiemanagement für mittelständische Unternehmen. Rationeller Energieeinsatz in der Praxis; Arbeitsschritte, Planungshilfen, Lösungsbeispiele. Dt. Wirtschaftsdienst, Köln

Weck M, Brecher C (2006a) Werkzeugmaschinen 4. Automatisierung von Maschinen und Anlagen, 6th edn. Springer, Berlin

Weck M, Brecher C (2006b) Werkzeugmaschinen Konstruktion und Berechnung. Konstruktion und Berechnung, 8th edn. Springer, Berlin. http://dx.doi.org/10.1007/3-540-30438-X

Weinert N (2010) Vorgehensweise für Planung und Betrieb energieeffizienter Produktionssysteme. Dissertation. Fraunhofer-Verlag, Stuttgart

Weinert N, Chiotellis S, Seliger G (2011) Methodology for planning and operating energy-efficient production systems. CIRP Ann—Manuf Technol 60:41–44

von Weizsäcker EU, Hargroves K, Smith MH, Desha C, Stasinopoulos P (2010) Faktor Fünf. Die Formel für nachhaltiges Wachstum, Droemer

Wiener N (2000) Cybernetics or control and communication in the animal and the machine, 2nd edn. 10. print. MIT Press, Cambridge

Wolfram F (1986) Aspekte der energetischen Bewertung von Produkten und Prozessen der Abtrenntechnik nach dem Prinzip der vergegenständlichten Energie. Dissertation

World Commission on Environment and Development (1987) Our common future, report of the World Commission on environment and development. Published as Annex to General Assembly document A/42/427, Development and International Co-operation: Environment. http://www.un-documents.net/wced-ocf.htm

Zein A, Li W, Herrmann C, Kara S (2011) Energy efficiency measures for the design and operation of machine tools: an axiomatic approach. In: Hesselbach J, Herrmann C (eds) Proceedings of the 18th CIRP international conference on life cycle engineering. Glocalized solutions for sustainability in manufacturing. Springer, Berlin, pp 274–279

Zhu J (2009) Quantitative models for performance evaluation and benchmarking. Data envelopment analysis with spreadsheets, 2nd edn. Springer, New York

Index

A. Zein, *Transition Towards Energy Efficient Machine Tools*, Sustainable Production, Life Cycle Engineering and Management, DOI: 10.1007/978-3-642-32247-1,

F (*cont.*)

G

I

L

M

N

O

P

Q